INTERPRETING PARABLES WITH PLANTS IN THEM

INTERPRETING PARABLES WITH PLANTS

Bringing home their meaning

Dr. Carolyn A. Roth

Illustrator

Maria Lin

Interpreting Parables with Plants in Them, Bringing home their meaning

Bibles Used:

Illustrator: internal plants: Maria Lin
Book cover: Margart Polinder at *Upsplash*

Publisher: Carolyn Roth Ministry Publishing, Roanoke, Virginia 24018 USA

Categories:
Religion and Spirituality>Christian Books and Bibles>
Bible Study and Reference>Bible Study>Guides

ISBN: 978-1-94919-19-9

Dedication

To my husband, Bruce Roth

Acknowledgements

Bruce Roth, my husband, helped edit the manuscript.

Maria Lin, designed and prepared prepared colored illustrations.

TABLE OF CONTENTS

INTRODUCTION

Purpose of Bible Parables

When Christian's think about parables, their minds default to Jesus' parables in the New Testament; however, parables didn't originate there. Parables are scattered throughout the Bible including in the Old Testament.

Seminarians and theologians debate what is and isn't a Bible parable. Often they deconstruct Bible allegories, metaphors, similes, and fables to determine if each meets criteria for a parable. In contrast to these worthwhile efforts, in this book a parable is a comparison where an image borrowed from the visible world is accompanied by a truth about the spiritual world and/or direction for spiritual living.

Parables go back to antiquity. In the ancient Middle East, parables were a common method of teaching. They have the following characteristics:

- Attract our attention. We want to know the punch line or end of the story.
- Stir up or excite us. Often, we remember a parable when other Bible content or a sermon is forgotten.
- Stimulate our thinking. We ponder and discuss a parable to understand its meaning. Through studying parables, we teach ourselves.
- Preserve truth. When we recall Inform our opinions. As we add life experiences and

maturity, we understand and apply parables at differing depth.

- a parable. i. e, of the wheat and the tares, we understand that when Jesus comes the second time, the righteous and unrighteous will be separated.

In both the Old and New Testaments, God's spiritual truths were illustrated using plants. Ancient peoples lived close to the land and nature. They depended on plants for food, both for themselves and for their livestock. They grew plants in gardens, i.e., herbs, grains, and grapes. Because ancient people understood plants, they readily interpreted parables illustrated by plants.

A single parable doesn't teach the entire Bible or gospel. It teaches a single point. The first step in understanding the point is to read the entire parable. Don't rush to judgement or conclude immediately that you understand its point. Remember, it is God's divine being and will that we interpret in a parable, not our own preferences. Second, study a parable's context. Review what happened before the parable was told (backstory). Know what occurred after the parable was given (outcome). The back story and outcome both aid parable interpretation.

Although individual parts of the parable develop its truth, not every part contributes to its interpretation. Some parts of a parable are there to make the story flow or make it easier to remember. For example, the number of virgins in the parable of the 10 virgins with olive oil lamps doesn't impact the parable's interpretation.

Although a parable has one primary interpretation, it can have many applications. When we apply a parable, we put it into practical use in our lives. Application is personal. My husband and I interpret a parable the same way; but we apply it differently. Bruce's application reflects his work as a skilled craftsman, while my application is to the church Bible garden.

How Plants Aid Parable Interpretation

God is infinite, eternal, everywhere, all knowing, and all powerful. In contrast, humankind lives a short life, i.e., 80 years on Earth. We are in a single place at any point in time. We don't know everything there is to know. We are relatively powerless. Despite the differences between God and humankind, God wants to communicate with us. In scores of Old and New Testament parables, God used plants to show his nature, spiritual truths about his kingdom, and how he wants us to live.

Interpreting Bible parables and plants in them isn't easy; 21st century society is far removed in time and culture from the Bible's original audience. The distance isn't only chronological, but social, political, and religious. Westernized individuals live in an industrial and technological-driven society. Often, contact with plants grown for food is second or third hand. Individuals don't have an "aha" moment where they understand a parable by its comparison to a plant, i.e., a fig tree, grapevine, or thistle.

To appreciate some parables, individuals, particularly children, need insights into plant life. What does the plant looks like, how does it grow, and how did biblical people use it? How does this plant's characteristics make it an appropriate, even optimal, illustration of God's character and will? Many 21st century children never sad growing plants. They need help with interpreting parables illustrated by plants.

The good news is that many basic plant characteristics—at least Bible plants—are fairly stable. This stability helps interpret plants in Bible parables. The current book centers on 20 parables illustrated by plants. I researched these parables and plants to learn why God chose each plant to illuminate a parable's spiritual truth.

Because Old and New Testament parables have different characteristics, I divided the book into a section on Old Testament and a section on New Testament parables. Often the same plants were used in Old Testament and New Testament parables. That's because Jesus and other New Testament writers borrowed heavily from Old Testament parables and illustrations.

The book is written for the average church member, rather than for seminarians and clergy. It's for individuals who want a different type of Bible study. This latter group may be individuals who want to grow plants in their home garden or in a church garden. They may be Master Gardeners, members of garden societies, and other nature clubs. Answer the study question integrated into each chapter. Questions can be pondered alone or discussed with friends in a Bible study group.

Each chapter contains a photograph of an original water-colored illustration of the plant used in the parable. Most Bible plants were identified by Jewish rabbis, historians, and Israeli botanists. When I cited specific plant names, I relied on opinions of Michael Zohary, Professor of Botany at The Hebrew University of Jerusalem, and of Nogah HaReubeni, creator of Neot Kedumim, the Biblical Landscape Reserve in Israel.

Section 1

OLD TESTAMENT PARABLES

The closest word to parable in the Hebrew language is *mâshâl*. *Mâshâl* translates: to resemble, become like, and compare. In a parable, Jewish writers compared the visible (physical) world to the spiritual world. Plant were almost everywhere and a central point of ancients' lives. That's one reason ancient writers often used plants in their parables..

The exact number of Old Testament parables is debated. One scholar identified as few as 10, while another identified as many as 26 parables.

Often scholars hold Old Testament parables to the same standard as those spoken by Jesus. They seem to forget or ignore that Israelite customs and language differed between Old and New Testament times. Further, Jesus was the son of God. His words and actions were superior to those of humankind.

In Old Testament parables, readers learn about the moral and social life in different periods in Israelite history. We see agricultural practices and religious conditions across time. Old Testament parables attract us by their truth about life. They show humankind as they were, not an idealized model.

Old Testament parables were spoken, seen in visions, and acted out by a cast of characters to include prophets, priests, kings, and an orphaned son. Many were focused locally (Samaria, Jerusalem) and related to a particular event; however, others had national focus, i.e., those of Isaiah and Ezekiel. At least one Old Testament parable was a fable where plants spoke; others were beautifully crafted poems.

While studying Old Testament parables, I noticed features which weren't common in Jesus' parables. First, some parables didn't stand alone; they included other actions. For example, Jotham's parable included a curse, while King Jehoash's included a warning. Second, many Old Testament parables were interpreted or explained through actions or outcomes in the same Bible chapter or subsequent Bible chapters. Only rarely was a parable's interpretation left to listeners or readers. Whether interpretations were or weren't provided, ancient listeners knew that they were to apply the parable to their lives.

The 10 parables reviewed in this section aren't an exhaustive list of Old Testament parables illustrated with plants. Some parables were excluded because both the plant and the message was discussed in another parable, i.e., some in the book of Ezekiel and one in the book of Daniel.

Chapter 1

BAALAM AND ALOE TREES

Getting Started: Read Numbers 24:1-9

The parable of the aloe and cedar trees was an oracle spoken by Balaam in Moab. Balaam wasn't an Israelites. Often the word *oracle* referred to a person, i.e., the oracle at Delphi; however, in this parable oracle meant an authoritative decision or opinion. The third oracle given by Balaam named two trees—the aloe and cedar.

Backstory

When Moabites saw Moses and the Children of Israel approaching their country, they were afraid that the "horde' was going to destroy the land, i.e., cut trees for firewood, consume pasture lands needed for their own livestock. Moabites wanted to turn the Israelites away. The problem was that Moab didn't have a standing army to fight against the Israelites. Until recently Moab was subjects of the Amorites. Moab was only freed from Amorite control when Israelites successfully defeated Sihon, the Amorite king.

The canny King Balak of Moab didn't have the military might to stop the Israelites from crossing his land. He decided to use guile to achieve his goal. To stop Israelites, King Balak sent for the most renowned seer/diviner in the known world—Balaam.

King Balak planned for Balaam to curse Israelites. In ancient times people believed that cursing a person or people could influence their outcome.

Balaam lived about 400 miles north of Moab near the Euphrates River. When Moab envoys first approached him, Balaam refused to accompany them to Moab because God told Balaam not to curse Israelites. Then, the Moabites sent a more distinguished delegation to Balaam and offered him more money. God allowed Balaam to go with Moabites; however, God warned Balaam that he could only speak the words God gave him.

Every Sunday school child knows what happened next. Balaam was riding a donkey toward Moab when suddenly the normally reliable donkey stopped in the middle of the path. The donkey refused to move forward despite Balaam hitting it with a stick. Then the donkey started to talk. Balaam's eyes were opened. He saw an angel blocking the donkey's pathway. The angel directed Balaam to go to Moab; but reminded him to speak only words God gave to him.

Parable

In Moab, Balaam spoke five separate oracles or parables directed at Israelites. Much to King Balak's disappointment, the parables didn't curse Israelites. In fact, Balaam's oracles blessed them! In his third parable, Balaam compared Israelites to aloe and cedar trees planted by God in watered land:

> How beautiful are your tents, O Jacob, your dwelling places, O Israel! Like valleys they spread out, like gardens beside a river, like aloes planted by the LORD, like cedars beside the waters. Water will flow from their buckets; their seed will have abundant water. Their king will be greater than Agag; their kingdom will be exalted (Numbers 24:5-7).

Characteristics of Aloe Trees

In this parable interpretation, the aloe rather than the cedar is emphasized. In contrast to the aloe of the New Testament, which came from an herbaceous plant, Old Testament aloe came from a tree. In the ancient near east, Old Testament aloe was from the eaglewood tree. Likely, Old Testament traders brought aloe wood from India. The Indian eaglewood is a massive evergreen that grows to 131 feet tall with a trunk (bole) of four-to-eight feet in diameter. Ancient Middle East people believed that the aloe tree was descended from a tree in the Garden of Eden. According to legend, Adam brought aloe tree shoots from Eden and planted them where he and Eve settled.

Aloe is made from agarwood in the eaglewood tree; but only about 10% of mature eaglewood trees produce agarwood. The fragrant oleoresin that permeates the heartwood of some eaglewood trees is produced by a fungal infection. Once the fungus establishes itself on the tree, the woody trunk turns a deep brown color. The darker the agarwood, the more valuable the aloe. Trees over 50 years produce the best agarwood. Agarwood is harvested, cut into small pieces, and burned. The result is a distinct aroma described as being a cross between sandalwood and balsam.

Because the eaglewood tree grew in India, likely Balaam never saw one growing in nature. When he included aloe trees in his parable, Balaam referred to the sweet smell of aloe in the eaglewood tree. His point was that the aroma of Israelites was attractive.

****

In the Bible, several places we read that Israelite sacrifices were a sweet aroma to God.

1. Read and summarize (don't just write them out) each verse about what a pleasing aroma to God is:

a. Genesis 8:21:

b. Exodus 29:18:

c. Leviticus 3:16:

2. How did sacrifices change from the Old Testament to post-Jesus New Testament times?

****

Meaning of Parable of Aloe Trees

Balaam's words more likely were about appearance of an eaglewood tree. Possibly, Balaam heard traders' descriptions of the majesty of aloe-producing trees. Something in or about the Israelite encampment resonated with descriptions Balaam heard of the uprightness of aloe trees.

The Bible story of King Balak recruiting Baalam to curse the Israelites occurred as the Israelites were making their final push into the Promised Land. If Balaam viewed the Israelite encampment immediately after the Israelites left Egypt, he wouldn't have compared them to upright aloe trees. As former slaves, their backs were bent, their heads bowed. In contrast, the current Israelites came to adult hood as free men and women. They stood upright. They lived in the dry, rugged Sinai. They were used to standing tall to look into the distance for the first sign of danger to their families or for signs of water and food.

Balaam was smart and knew his craft well. He knew Israelite history. He knew how they left Egypt. He knew about their 40-

Aloe Tree
AGARWOOD
Maria Lin

year trek in the wilderness. Possibly, Balaam compared the 50 years it took the eaglewood tree to produce mature agarwood (aloes) with the 40 years the Israelites small number of aloe trees convert to agarwood and sweet-smelling aloes. So too, the children of Israel was a small number of aloe trees convert to agarwood and sweet-smelling aloe. So too the Children of Israel was a small number, or percentage, of the world's population. Yet, this small nation was unique in being God's chosen people; they were a sweet aroma to God.

Finally, Balaam may have known that aloe trees (eaglewood trees) were adapted to a variety of environments, i.e., rock, sand, well-drained slopes and ridges, and land near swamps. Similarly, Israelites adapted to a variety of living environments, i.e., Canaan, the Nile River delta, the Sinai Desert, and now Moab.

****

Think about your national culture and society

1. Have you seen any events and happenings declared okay or permissible in society that God declared sinful, damaging, or depraved? How are you, or could you be, a sweet aroma to God in these situations?

2. List at least three actions – changes in your life – that would make you more fragrant to God.

a.

b.

c.

3. As I have aged, I am becoming less adaptable. Would you describe yourself as adaptable? Should age be a factor in adaptability?

****

The spiritual interpretation of Balaam's parable is: what God has declared blessed, man shouldn't curse. The opposite is also true, what God has declared as sinful, damaged, and depraved, man can't declare as valuable and good. Perhaps an application is that we are to grow and fulfill our potential in whatever environment God sets us; like agarwood (aloe), we should be a sweet aroma to God.

Outcome

If we leave the story of Balaam at this point, we could conclude that Balaam wasn't such a bad guy. After all, he followed God's direction and didn't curse Israelites, But, there is more to the story of Balaam. The Bible recorded that after giving a fifth oracle, Balaam returned home. Possibly, he went back north; but Balaam didn't stay there. The Israelites killed Balaam along with five Midian kings in a battle northeast of the Dead Sea. The Israelites were justified in killing him. Balaam was responsible for advising Moabites and Midianites to use women to seduce Israelite men to sexual immorality and idol worship. This advice led to the slaughter of over 24,000 Israelites.

Takeaway

Only 10% of eaglewood trees produce sweet-smelling agarwood (aloe). Are you producing a sweat-smelling aroma for Christ?

Chapter 2

TREES CROWN A KING

Getting Started: Read Judges chapter 9

Technically, the parable of the trees crowning a king is a fable because plants don't speak. In the ancient near east, people were familiar with fables. Hearers knew the fictitious story had a moral that applied to everyday life. Often, they discussed, even argued, the story to figure out its interpretation and application.

Backstory

Actions of the Israelite judge, Gideon, were important to this parable. Called by God, Gideon was from the tribe of Manasseh; his home was in the Jezreel Valley. Gideon led the Israelites to throw off the seven-year tyranny of the maundering Midianites. One of the many positive characteristics of Gideon was his refusal to be king over the Israelites after he defeated them. His words were, "I will not rule over you, nor will my son rule over you. The Lord will rule over you" (Judges 8:23). Gideon had 70 sons by his wives and one son, Abimelech, by his concubine. Abimelech lived with his mother's people in Shechem. Abimelech's name means "my father is king."

After Gideon's death, Abimelech negotiated with his uncles and men of Shechem to crown him king. Abimelech and a group of paid adventurers murdered Gideon's legitimate sons with the exception of the youngest, Jotham. Abimelech's coronation as king occurred at the great oak tree at Shechem. At this site Joshua had made a covenant between God and the Israelites. He recorded God's decrees and laws in the Book of the Law of God. In the approximate 250 years between Joshua and Gideon's deaths, these Israelites moved from belief that God was their king to putting their confidence in an earthly king.

****

Before Israelites entered the Promised Land, God instructed them to kill all Canaanites.

1. How could the situation among Jotham-Abimelech-Shechemites been avoided?

2. Identify times when you blatantly disobeyed God's instructions. How did those times work out for you?

3. Identify what you would or could do differently. Don't you wish hindsight was foresight?

****

When Jotham gave this parable, he was heart sick. Not only had his father, the stable person in his life, died; but his half-brother murdered every other brother. Although the Bible doesn't say so, likely Abimelech seized all of Gideon and his brothers' lands. Jotham, legitimate son of Gideon, was left with nothing. On top of everything else, the men of Shechem and surrounding areas proclaimed Abimelech king! Jotham was outraged by the injustice perpetrated by Abimelech and the same people Gideon fought to free from the rapacious Midianites.

A Parable Plus

On the day that Abimelech was crowned, Jotham climbed to the top of Mount Gerizim and loudly proclaimed this parable:

> Listen to me, citizens of Shechem, so that God may listen to you. One day the trees went out to anoint a king for themselves. They said to the olive tree Be our king. But the olive tree answered, Should I give up my oil, by which both gods and humans are honored, to hold sway over the trees? Next, the trees said to the fig tree, Come and be our king. But the fig tree replied, Should I give up my fruit, so good and sweet, to hold sway over the trees? Then the trees said to the vine, Come and be our king. But the vine answered, Should I give up my wine, which cheers both gods and humans, to hold sway over the trees? Finally all the trees said to the thornbush, Come and be king. The thornbush said to the trees, If you really want to anoint me king over you, come and take refuge in my shade; but if not, then let fire come out of the thornbush and consume the cedars of Lebanon! (Judges 9:7-15).

Jotham didn't stop with a parable, he included a curse. Jotham's curse was that if the people acted honorably toward Gideon who fought for them and rescued them from the Midianites, then Jotham wished them joy in their new king. However, if they acted dishonorably, then let fire come out of Abimelech and consume the citizens of Shechem and Beth Millo and let fire come out of the two towns and consume Abimelech.

After giving the parable and curse, Jotham fled south to Beersheba. Jotham made his home at Beersheba because he feared Abimelech.

****

Have you ever wondered how Jotham's life turned out?

1. Read Romans 3:23. How does Romans 6:23 apply to Jotham's life?

2. Do you believe that Jotham ignored any of God's instructions? Did he deserve to be separated from Gideon's family? Do we not ever deserve what we get in life?

3. What is your answer to the age-old question of why bad things happen to good people?

****

Characteristics of Thornbushes

The thornbush was known commonly as the *atad.* It is larger than all, or most, fruit trees native to Israel. In the wild, thornbushes can grow 30–35 feet tall; consequently, it grows as a tree more often than a bush.

Branches are dense and intertwined. Sometimes, branches dropped downward, almost touching the ground. When this happens, sunlight can't reach the area beneath the bush and little grows there. Each leaf has a pair of thorns. One hard thorn is straight, while the other is hooked.

The thornbush root system is deep and wide spreading. Frequently, thornbush roots leach all nourishment from the surrounding soil. For fruit trees, i.e., olive, fig, pomegranate, to grow successfully, farmers must first remove all thornbushes before planting fruit trees.

In Israel, thornbushes produce flowers between March and October. The fruit is yellow and small, less than one inch in diameter. Each fruit contains a large stone (pit) surrounded by fleshy pulp. Fruit is best eaten green. In no way does the quality of the thornbush's fruit compare favorably to the olive, fig, or grape. In ancient Israel, thornbush fruit was only eaten by people who were very poor or in time of famine The small fruit tastes similar to an unripe apple.

Meaning of Parable (Fable)

Overall, Jotham's choice to use the thornbush as a symbol of Abimelech's lack of qualifications to be king demonstrated that he knew how trees grew. The physical interpretation of this fable/parable was that a) Abimelech was a lesser man than Gideon, b) Abimelech's reign would be disastrous for the Israelites whom he ruled, and c) fruit of Abimelech's reign would be less than the present situation where the Israelites were ruled by God and his appointed judges. The thornbush agreed to be king of the trees with the caveat that other trees rest in its shade. Resting in a king's shade was a common euphemism for a king caring for the people he ruled and for the people submitting to the king's will.

The thornbush threatened if the other trees refused to rest in its shade, then its fire would consume the cedars of Lebanon. Essentially, the thornbush threatened that if it was made king, and then subjects refused support, it would destroy them. In Jotham's parable, Abimelech was the thornbush who agreed to be king after all honorable men declined. The cedars of Lebanon symbolized the Israelite people over whom Abimelech ruled.

Although Abimelech was king, his subjects wouldn't be able to grow under his kingship. Abimelech would leach resources from his subjects in much the same way that the thornbush leached nutrients from soil. Jotham knew that other trees and smaller plants couldn't grow under thornbushes.

Thorn tree
Maria Lin

A spiritual interpretation of Jotham's parable and this horrific situation is that when we move ahead of God's timing, the outcome is sub-optimal, if not disastrous. God's choice for the first human king of the Israelites, King Saul, wasn't crowned for another 70 plus years. At this time (immediately post Gideon) in Israelite history, God's plan was for Israelites to have judges as leaders. When the Israelites went their own way and crowned a king, the result was inferior to God's plan for them.

Outcome

After three years of Abimelech's kingship, God sent an evil spirit between Abimelech and the Shechemites. We aren't sure how the evil spirit worked. Perhaps, after becoming king, Abimelech had time to think about his status. True, he was undisputed king because he killed all opposition; however, his backers were mercenaries. Further, the townsmen who supported him were the type of men who condoned the murder of Gideon's sons. How secure was his throne? How likely was it that his sons would succeed him as king?

****

Why did Abimelech want to be king?

1. Did Abimelech's love of power separate him from God or was some other sin occurring in his life? If another sin, what do you think it was?

2. God had a plan for Abimelech's life which Abimelech negated by his actions. What are some alternate ways that Abimelech could have acted, along with possible outcomes?

3. Read Romans 6:23. How does Romans 6:23 apply to Abimelech's life?

****

Townsmen over whom King Abimelech ruled may have had second thoughts. They remembered Gideon's words that God should rule over them, not a man. They couldn't trust a man who slaughtered his 70 brothers. Some Shechemites ambushed and robbed people who passed by Shechem, thus, thumbing their nose at Abimelech and his "power" to keep the peace.

A man, Gaal, and his brothers moved into Shechem. He was a Canaanite. Many townsmen turned from King Abimelech and started to follow Gaal. While Gaal and his followers were eating and drinking, they cursed Abimelech. Abimelech heard about their disrespect and attacked Gaal and Shechem. Gaal was ousted from Shechem; however, Abimelech proceeded to attack Shechem. About 1000 men and women took refuge in the Shechem tower. Abimelech ordered his men to gather branches from Mount Zalmon. He piled branches against the tower and lit them with fire. All 1000 people in the tower died from the fire, heat, and smoke inhalation. The Bible didn't identify the types of trees or branches used to burn the Shechem tower; but, likely the thornbush was included among them.

Not satisfied with destroying Shechem, King Abimelech and his army went to Thebez and captured the town. He attacked Thebez's tower. A woman inside the tower dropped a millstone that cracked Abimelech's skull. Abimelech didn't want it known that a woman killed him, so he persuaded his servant to kill him. When Abimelech's followers saw that he was dead, they went home; thus, ended the short three-to-five- year reign of the first king of Israel.

Jotham's curse came true. Fire came out of the thornbush (Abimelech) and destroyed the citizens who supported him. Out of the towns Abimelech ruled, came fire—in this case rebellion—that caused King Abimelech's death.

Takeaway

Do you think that Jotham's curse had an impact on the outcome of the story?

Chapter 3

TRANSIENT PLANTS / WICKED MEN

Getting Started: Read Psalm 37

Some Old Testament parables tell a complete story, i.e., trees crowning a king. Others are short; sometimes terse. Whether long or short, each parable has a spiritual message that unfolds through ideas, events, or natural objects in the physical world. In Psalm 37, King David wrote three parables of one-to-two verses each. These short parables are about wicked, ruthless men.

Backstory

As we read Psalm 37, we imagine an older and wiser King David. He isn't any longer the brash aspirant to Israel's throne or a newly crowned king. This King David has seen a wide range of events and people in his lifetime. He acknowledged his sin of having Uriah killed so he could marry Bathsheba. He knew his daughter, Tamar, was raped, and subsequently dealt with the murder of Crown Prince Amnon. King David was deposed as Israel's king and fought a heart-breaking battle to regain his throne. God, who David adored, told David that his hands were too bloody to build God's temple.

King David wrote psalms before and after he became King of Israel. Many were written to praise God or were prayers to God.

In contrast, Psalm 37 is directed toward all who will listen. David wrote this psalm to praise God and to teach individuals who read or heard it. In Psalm 37, some of the 40 verses are discrete thoughts while others are two-verse couplets. All are loosely organized around the theme that wicked men perish while good men prosper.

David had some friends and many loyal followers; but, over his lifetime he experienced plots from many enemies. In Psalm 37 he told righteous men "Do not fret because of evil men or be envious of those who do wrong; for like the grass they will soon wither, like green plants they will soon die away" (Psalm 37:1-2). David was very clear about the outcome of wicked men "But the wicked will perish: The Lord's enemies will be like the beauty of the fields, they will vanish" (Psalm 37:20).

Parable of Green Tree

Here, I focus on David's third plant parable in Psalm 37. David wrote:

> I have seen a wicked, ruthless man spreading himself like a green laurel tree. But he passed away, and behold, he was no more; though I sought him, he could not be found" (Psalm 37:35-36 ESV).

****

When you see wickedness and ruthlessness actions,
Can you identify them?

1. Define each word:

    Wickedness:

    Ruthlessness:

2. How are these words similar and how do they differ?

3. Mentally, identify who in your circle of friends is wicked? Who is ruthless?

****

Characteristics of Laurel Trees

David's green bay tree is known to Americans as the bay laurel tree. Characteristics of the bay laurel tree make it a fitting comparison to the transience of wicked, ruthless men. In Israel, laurel trees are called the sweet bay laurel because bay leaves come from the tree. In America bay leaves are considered a spice and are purchased as small, dried leaves.

During David's reign laurel trees grew wild from Mount Hermon in northern Israel southward through the Jezreel Valley and into the Judean Mountains. Today most laurel trees are 6-12 feet tall; however, in ancient Israel laurel trees grew to 60 feet. In Israel the laurel tree is an evergreen, retaining green leaves year around. A way to identify a laurel leaf is to bruise or cut it and smell the sweet aroma; the aroma is of a bay leaf.

One of the most important attributes of laurel trees—and one that King David apparently knew—was that laurel trees thrive where they are planted. They tend to wilt and even die if they are moved repeatedly.

Ideally, gardeners plant laurel trees and allow them to grow in place. Laurel trees are sensitive to cold and frost. Bay laurel trees thrive in temperatures between 60°F - 75°F (15° - 24°C). During active growth, the ideal range narrows to 65°F - 70°F (18° - 21°C).

Laurel branch
Maria Lin

Meaning of Parable

In this parable King David compared a man to a flourishing tree. The man's activities, although wicked and ruthless, were successful. Green laurel trees grow and spread. So does a wicked man. He spreads beyond himself, his family, his community, and sometimes even his city and country.

The wicked man's actions were cruel and without pity. He didn't care how he achieved wealth and influence. This is the man who took property from widows and orphans, but, and this "but" is reassuring. This merciless man disappeared. He wasn't there when King David went looking for him.

The parable doesn't tell the reader if man died or if his wickedness caught up with him. Perhaps, someone even more wicked brought him down. King David's parable tells readers only that the man "was no more." David even looked for the wicked man. Perhaps, David wanted to know if he was still wielding his power, but David couldn't find him.

David concluded that wicked men don't endure; they have no staying power. Perhaps, like a laurel tree, wicked men can't tolerate adversity—to include cold weather, frost, and transplanting. The laurel tree and a wicked man flourished, but, only in a narrow environment.

****

Ponder whether or not you trust God with events in your life.

1. If you trust God, how does your life show it?

2. If you do not trust God, how does your life show it?

3. Any changes you want to make in how you act?

****

Outcome

In the three parables in Psalm 37, King David went beyond identifying the puzzle of seeing wicked individuals prosper. David advised not to fret when evil men (and women) prosper because fretting leads to evil. When David said evil, he meant anger, resentment, or mimicking wicked and ruthless men's business practices. Instead, we should refrain from anger and hope in the Lord. Hoping in the Lord means we shoulder bad as well as good times from God. We accept cares and joy.

****

Are there alternatives to anger and resentment at seeing evil and ruthless individuals prosper? What are some of these alternative that you could exhibit?

****

Takeaway

When King David wrote that he saw wicked and ruthless men who flourished like a green laurel tree in its native soil, he was thinking of a mature laurel tree with a broad canopy and numerous branches. Prosperous, wicked men never seemed to have setbacks; but we, even though we are Christians, have setback.

When we concentrate on flourishing in the garden where God plants us, we obsess less about whether or not ruthless people get ahead.

Chapter 4

VINEYARD PARABLES

Getting Started: Read Psalm 80:8-13 and Isaiah 5:1-6

Do you want your heart to break? Read the next two parables. Learn about Israelites' rejection of God. Although written at two different times in Israelite history and by two different men, both parables are a cry of anguish from God to his chosen people. Both identify God as a gardener and Israelite as his vineyard. A vineyard and vine are synonyms with the nation of Israel (both Northern and Southern Kingdoms).

Backstory

Parts of Canaan were ideally suited for vineyards and grape production. A single cluster of grapes could weigh 24 pounds. Remember, the spies sent by Moses into Canaan (Number 13:23)? They returned with one cluster of grapes that required two men to carry.

Grapes became a principle export of Israel under King Solomon. Hillsides, where wheat didn't grow well and cattle couldn't be pastured, were ideal sites for vineyards. In the Northern Kingdom, the hill country in Manasseh and Ephraim produced grapes and grape production occurred in Naphtali in the far north of Israel. In Judah, the area around Hebron, with its many hill, was perfect for vineyards. In En Gedi, vineyards were plentiful.

After the death of King Solomon, the Israelite kingdom divided. The Northern Kingdom, called Israel, included 10 tribes: Ephraim, Manasseh, Dan, Zebulun, Asher, Naphtali, Issachar, Gad, Reuben, and Simeon. Some of the greatest Israelite prophets, Elijah and Elisha, lived in the Northern Kingdom.

The Southern Kingdom was known as Judah. At times, Judah included both the tribes of Judah and Benjamin. At other times, Benjamin seemed more aligned with the Northern Kingdom. Levite towns were located in both Israel and Judah; but, for the most part the Levites were associated with Judah and the Jerusalem Temple.

The first Northern Kingdom king introduced idol worship, which was embraced by much of the population. God sent prophets, i.e., Elijah, Elisha, Amos, Hosea, to warn Israel that their idol worship and flaunt of God's laws would cause God to reject them.

Bible scholars believe that Psalm 80, a lament, was written in response to the fall of the Northern Kingdom. The psalmist was from the family of Asaph, a Levite singer, poet, and head of King David's choir. Likely, Psalm 80 was written in Jerusalem from where the Northern Kingdom collapse was closely observed:

In the Southern Kingdom, Judah, descendants of King David ruled until 586 BC. Despite prophets, i.e., Isaiah, Jeremiah, Ezekiel, warnings and calls for repentance, Judah turned from God to worship a myriad of idols. Jerusalem fell to the onslaught of the Babylonians. Many citizens not killed were taken captive to Babylon.

****

Both the Northern and Southern Kingdoms forgot that God was their hedge. They believed their armies protected them.

1. What do you believe is your protection from the world? Money? Gated community? Job? etc.

2. Overall, what do we as Americans believe are protections against foreign oppressors? How are our beliefs similar and different from antient Israelites?

3. How could you increase your life's security?

****

Two Parables

Asaph's parable lamenting the downfall of the Northern Kingdom reads:

> You transplanted a vine from Egypt; you drove out the nations and planted it. You cleared the ground for it, and it took root and filled the land. The mountains were covered with its shade, the mighty cedars with its branches. Its branches reached as far as the Sea, its shoots as far as the River. Why have you broken down its walls so that all who pass by pick its grapes? Boars from the forest ravage it, and insects from the fields feed on it (Psalm 80:8-13.)

Isaiah's parable used a similar comparison of a gardener who planted a vineyard. Instead of lamenting the downfall of the Northern Kingdom, Isaiah identified the fate of Judah. The song of the vineyard is a love song that described God's love and care for his people, how they rejected him, and ultimately God's punishment of Judah.

> I will sing for the one I love a song about his vineyard: My loved one had a vineyard on a fertile hillside. He dug it up and cleared it of stones and planted it with the choicest vines. He built a watchtower in it and cut out a winepress as well. Then he looked for a crop of good grapes, but it yielded only bad fruit. "Now you dwellers in Jerusalem and people of Judah, judge between me and my vineyard. What more could have been done for my vineyard than I have done for it? When I looked for good grapes, why did it yield only bad?
>
> Now I will tell you what I am going to do to my vineyard: I will take away its hedge, and it will be destroyed; I will break down its wall, and it will be trampled. I will make it a wasteland, neither pruned nor cultivated, and briers and thorns will grow there. I will command the clouds not to rain on it (Isaiah 5:1-6)

Characteristics of Vineyard Fences

Both the psalmist and Isaiah referenced vineyard fences. The psalmist asked God why he broke down vineyard walls and removed its hedge so that all who passed could pick the grapes. Further, boars from the forest could ravage the vineyard. Isaiah prophesied that if Judah didn't repent, God would remove their walls, and the country would be trampled. Briers and thorns would grow in former vineyards.

Usually, ancient Israelites enclosed vineyards with fences. Often farmers dug a ditch around the vineyard. The earth from the ditch was thrown to the inner side of the ditch. Then, fence posts and thorny plants were placed on the berm. At other times, a wall of stones or sun-dried mud took the place of the earthen fence with its thorny plants. Boulders and rocks were plentiful in

Vineyard
Maria Lin

Canaan. Large boulders were removed from the vineyard area and used for stone fences. Small stones weren't removed from the soil; their presence encouraged the soil to retain moisture, needed by vineyard's soil.

God built a watch tower to protect his vineyard Israel. For centuries Israelite vineyards had booths and watchmen who protected the vineyard from marauders. The booth became the watchman's home for the summer months. Sometimes a simple booth was constructed on a high spot where the watchman could view the entire vineyard.

Frequently, watchman's booth was made of tree branches and boughs and provided shelter from the rays of the sun. At other times a more durable tower was constructed, especially if the watchman's family lived with him during the summer months. These towers were from 10-40 feet tall. In the winter months the watch tower was deserted. The watchmen that God set over his vineyard, Israel, were first judges and later kings; even prophets were the watchmen. Nonetheless, Israel was a theocracy with God as their ultimate watchman and ruler. Israelites' first allegiance was to be to God.

Meaning of Two Parables

In Isaiah's parable the vineyard was planted with the choicest vine. Farmers traded for the best vines available or purchased the highest quality vines that they could afford. As the vine grower devoted himself to the vineyard and grapes, God devoted himself to the children of Israel. He did everything possible to select good vines, protect them, and promote good growth in them.

****

God judges our actions

1. Imagine God is viewing your life's actions (and he is). What does he see doing right?

2. Imagine God is viewing your life (and he is), what does he see you doing inconsistent with his statutes?

3. Is your life in the correct righteous-unrighteous proportions? What could you change to improve this proportion?

****

The spiritual interpretation for both Psalm 80 and Isaiah 5:1-6 is that there are consequences to sin and both the Northern and Southern Kingdoms would experience those consequences. The consequence of Israel and Judah's sins was that God abandoning them. God removed his protection from both kingdoms with the result that both were attack and ruined. In their pride both kingdoms forgot that God was their hedge. They believed that armies were protection for their borders and for their populations.

Outcome

Although God anticipated a yield of good grapes from his vineyard, he got only bad grapes. A Northern Kingdom prophet summed up what God wanted from his people: for them to act justly, love mercy, and walk humbly with him. God expected a yield of righteousness. Instead, the people were rebellious and idolatrous.

In the Southern Kingdom, Judah, descendants of King David ruled until 586 BC. Despite prophets, i.e., Isaiah, Jeremiah, Ezekiel, warning and call for repentance, Judah turned from God to worship a myriad of idols. Jerusalem fell to the onslaught of the Babylonians. Many citizens not killed were taken captive to Babylon.

In these two parables we can learn that God's judgment is proportional. After becoming a nation separate from Judah, immediately the Northern Kingdom (Israel) turned to idol worship, i.e., Jeroboam I set a golden bull in Dan and Bethel and told Israel's citizens that the bulls were the gods that brought them out of Egypt. Over the next 200 years, most of the nation's population repudiated God. In response, God allowed the nation of Israel to fall and most of its people to be exiled from the land.

Many of Judah's leaders and population rejected God through idol worship, but, some kings, i.e., Hezekiah and Josiah, initiated reform and returned to worship of God. Although God removed his protection from Judah and Jerusalem, he provided a way for these Jews to return to Jerusalem, rebuild the temple, and reestablish themselves as a nation.

Takeaway

God was so disturbed by the actions of citizen of both the Northern and Southern Kingdoms of Israelites, that he asked: "What more could have been done for my vineyard than I have done for it?" (Isaiah 5: 4).

God asks you and the Church what more can he do for you?

Chapter 5

GOD, A LIVING TREE

Getting Started: Read Hosea 14:8

The parable of the green tree is one of the Bible's miniature parables. Not only is it brief, but it tends to be obscure. Obscure parables allow the Holy Spirit to call forth prayerful study, so parables are fully interpreted. When Hosea spoke the parable of the evergreen tree, it stimulated the ears of the people who heard it. Hopefully, the parable also made people think about its interpretation. Today, we read the parable and visualize the majesty of a green tree.

Backstory

Hosea was a prophet to the Northern Kingdom (Israel). Hosea lived and prophesized during the middle of the eighth century BC, during the final tumultuous days of Israel when six kings reigned in 25 years. Most rulers were murdered by their successors with only one effective transition from father to son. The major tribe of the 10 Northern tribes of Israel was Ephraim. Frequently, Hosea referred to the Northern Kingdom as Ephraim.

Generally, Christians remember Hosea because God directed him to marry an adulterous woman, Gomer. Many scholars consider Hosea and Gomer's story a parable in and of itself.

Hosea's interactions with Gomer are contained in the first three chapters of his book. In the other 11 chapters, Hosea implored the Northern Kingdom to repent so that God could heal their waywardness.

****

Idols are what we place ahead of God.

1. What idols do you worship in your life, i.e., spouse, children, money, career?

2. In your Bible, read Exodus 20:3-4 and 1 Corinthians 10:14. What did God command about idols?

3. How can you lessen your own idol worship? What steps can you take?

****

Hosea averred that Israel's disloyalty to God and idol worship was spiritual adultery. Hosea identified Assyria as the source of God's judgment on the Northern Kingdom. Because Hosea came from the Northern Kingdom, he knew every pride and perversion of royalty and citizen alike. Yet, Hosea spoke of God's love, mercy, and forgiveness.

In the first three verses of Hosea chapter 14, Hosea exhorted Israel to do three things:

- stop sinning by worshiping idols made by their hands
- ask God to forgive their sins
- return to worshiping God.

Hosea assured Israel that Assyria can't save them, nor can war horses (from Egypt). God alone can save Israel.

****

The title of Hosea chapter 14 is: "Repentance to bring Blessing"

1. If you are as old as I am, you heard the saying "Love means that you never have to say you are sorry." If the statement is true, then I never have to tell God I'm sorry. Do you think that statement is accurate?

2. Look up the definition of repentance in the dictionary. Is repentance more than telling God you're sorry?

3. How does repentance bring blessings?

****

The next four verses (Hosea 14:4-7) are a beautiful outline of how God will heal Israel's waywardness if the nation returns to God. When we read the verses, our hearts rejoice at what God – even at this late time – is willing to do for Israel. God's healing is expressed using plant metaphors:

- God will be like the dew and Israel will blossom like a lily
- Like a cedar of Lebanon, Israel will send down roots and send up new growth
- Israel's splendor will be like an olive tree
- Israel's fragrance will be like a cedar and men will again live in its shelter
- Israel will flourish like grain and blossom like the grape vine
- Israel's fame will be like wine from Lebanon.

Parable of Green Cypress Tree

After assuring the Northern Kingdom, that God could heal Israel, Hosea offered a parable. Initially, I read the parable in the New International Version (NIV) Study Bible; however, I disagreed with the NIV translation of the tree as a pine tree. The pine tree that grows in Israel is the Aleppo (Syria) or Jerusalem pine. Although an evergreen, normally the Aleppo pine is short-lived (70-100 years). It has a shallow root system. The pine tree is a poor example, and perhaps a poorer translation, of the tree in Hosea 14:8.

Other Bible translations identified the tree as a cypress tree (AMP, RSV). Characteristics of the cypress make it a more likely candidate for a comparison with God. Here is the parable:

> "O E′phraim, what have I to do with idols? It is I who answer and look after you. I am like an evergreen cypress, from me comes your fruit" (Hosea 14:8 RSV).

Characteristics of Cypress Trees

The wild cypress tree of Israel was commonly called the Mediterranean cypress. The cypress remains green year around. The cypress tree is large, strong, durable, and full of resin which acts as a barrier against water seeping into the wood. Cypress trees are enduring, typically living at least 1000 years. In Genesis a cypress-wood ark carried eight individuals through the raging flood that destroyed life in the then known world. Ancient Middle Eastern peoples viewed the cypress as a symbol of immortality.

On the Mediterranean cypress, loosely hanging branches grow from the central column in a whorl. Branches are rounded and have small, elongated needles for leaves. Cypress tree branches and stems grow close together. These growth patterns plus year-around needles means that people, who take protection under a cypress tree, have little exposure to rain or snow. Likewise, the thickness of branches means travelers who rest under the cypress are rarely exposed to the sun. What a wonderful analogy of how God wanted to protect Israel, i.e., from the sun, rain, and snow.

Instead of brightly blooming flowers, the cypress tree produces cones. The same tree produces both male and female cones. On the Mediterranean cypress, pollination of female cones occurs in the summer. After pollination, the female cone's green ovules turn to seed. Usually, fertile cones turn brown and drop to the ground the year following pollination.

Meaning of Cypress Tree Parable

In the Old Testament God revealed in many places and ways that he wanted to protect his chosen people and make their lives better, i.e., lead them in green pastures and beside quiet waters, heal, guide, restore, and comfort them, and give them the kingdom forever and ever.

Perhaps, nowhere is God's caring so forthrightly and succinctly presented as in Hosea 14:8. He told Israel that he, not an idol, answers them and looks after them. He is like a green cypress tree. From God comes Israel's fruit, both their food and their righteousness. Not surprisingly, the cypress tree was associated with immortality through God's divine will.

In Hosea, God liken himself to something alive, as he is alive. That living organism was a tree with a lovely smell and which was disease resistant. Although ancient people used the cypress tree to symbolize immortality, God doesn't just symbolize immortality; he is immortal. By likening himself to a green cypress tree, God compared himself to a tree that lived so many more years than a human lifetime.

Cypress, green tree
Marca Lin

The parable's spiritual interpretation is that human-made idols aren't immortal; they aren't alive. They are statues, images, man's creations. Some have ears but they can't hear. Some have mouths but they can't speak. Having a head is not the same as having a brain and mind.

Immortality, including long life for an individual or a nation, comes only from God. God promised that if Israel would follow him, the country would once again be fertile, green and grow.

I read through the Bible several time; but, perhaps, unlike you, I never paid attention to Hosea. Hosea is a relatively murky book which other than the parts about Hosea and Gomer doesn't seem all that intriguing. But Hosea 14:8 is the only place (that I know of) where God compared himself to a living organism. We know that God is enduring like the mountains, the soil, and the ocean.

Outcome

In the Northern Kingdom, Hosea's call for the people to repent was ignored. After 200 years (722/721 BC) Assyria conquered the Northern Kingdom. Many inhabitants were exiled throughout the Assyrian Empire.

Takeaway

Evergreen trees are alive. God is alive. Probably, you have often heard that we will all have immortality. The question is where each of us will spend that immortality: in heaven or in hell. Read the following verses, then answer the question of where you will be spending eternity: Ephesians 2:8-9, Acts 4:12, and Romans 10:9-10.

Chapter 6

THORNS & FOOLS

Getting Started: Read Ecclesiastes 7:6

When Bible-readers were asked what they learned from reading Ecclesiastes, most answered with some variation of, "meaningless, meaningless, all of life is meaningless." With that answer, some proceeded to identify another Bible passage that told a more compelling story, stated a clearer principle for a Godly life, or prophesized an exciting end-time event. Yet, the Teacher's sometimes sly and thinly perceived wisdom gives direction for life to individuals who want to learn wisdom.

The parable of crackling thorns is short, contained in one verse. The crackling thorn parable was meant to be experienced by the ears. Indeed, we hear both crackling thorns as they burn and the loud boisterous laughter of fools.

Backstory

Ecclesiastes is part of the wisdom literature of the Hebrew people. Bible scholars aren't sure who wrote Ecclesiastes. Some suggested that Solomon authored the book. Others proposed that a citizen rather than a king was the writer and that Ecclesiastes was written long after the reign of Solomon. The writer referred to himself as "Teacher" (Ecclesiastes 1:1).

The focus of Ecclesiastes is that individuals should enjoy the life that God gives them. The author questions how mankind can live meaningfully, purposefully, and joyfully in the relatively short time allocated to each individual. Simultaneously, the Teacher answers his question by saying that a full life can be had if individuals place God at the center of activities, i.e., work, family, church. The Teacher concluded that life under God was good, despite its multiple puzzles, paradoxes, and conundrums.

****

Appeal

Name three things about a non-Christian lifestyle appeal to you.

1.

2.

3.

Do you want a Christian lifestyle? Why or why not?

****

Parable of Crackling Thorns

The parable of crackling thorns is included in a chapter titled: The Contrast of Wisdom and Folly (ESV). Despite the chapter's name, the parable addressed folly, rather than wisdom. The parable reads:

> Like the crackling of thorns under the pot, so is the laughter of fools. This too is meaningless (Ecclesiastes 7:6).

Thorny burnet
Maria Lin

When the Teacher concluded the parable with, "This too is meaningless," he linked the laughter of fools to burning thornbushes. A closer look at the word "meaningless" showed that it means "futile" or "has no purpose." Synonyms are "pointless and ineffective;" while the antonym is "useful."

Thorny Burnet Fueled Cooking Fires

The thorn plant in Ecclesiastes 7:6 is the thorny burnet commonly called the prickly burnet. The Hebrew word for crackling is *qôw*l (*qôl*), an unused root word meaning a voice or sound. In present-day Israel, thorny burnet can be found from northern Mount Hermon through Eilat, in the far southern desert. Like fools and their laughter, thorny burnet is widespread.

Two, when individuals use thorny burnet plants as cooking fuel, three plant characteristics are important: First, although the burnet plant grows in dense clumps, it isn't a good source of cooking heat. Dry thornbushes burn rapidly. Thorn heat flairs up quickly, but the fire dies down just as quickly. Thornbushes don't provide sustained heat to cook a meal or bake bread as heat generated from wood. Similarly, the laughter of a fool occurs readily; but can end as reality sets in, the amusing event passes, or the joke is over.

Three, in Israel thorny burnet flowers bloom at the end of the rainy season, February–April or earlier in Israel. Mature female flowers are set in four-to-five joined petals. The top of the petals are covered tips of sepals (modified leaves set around the petals) which makes the female flower resemble a covered pot. These flower "pots" are numerous. In a fire, flower "pots" produce a small explosion.

Meaning of Thorn Fire

This parable that the Teacher wrote compared a fire of crackling thorns to the laughter of fools. In ancient Israel, cooking fires were in outer courtyards or inside homes. The time of year determined where the fire was located. In hot weather it was in the courtyard. In cooler or cold weather, the fire was located inside the home to add warmth. Many poor Israelites cooked over a simple hole in the ground (a fire pit) with rocks around it. Others had ovens, either inside the home or in the courtyard. Sources of fuel included wood, animal dung, and thorn bushes.

****

How could you act less foolishly and more wisely?

1. Where or how would you start this process?

2. What would you have to give up?

3. What would you have to or want to add to your already busy life?

****

Although a thorny burnet fire may snap, crackle, and pop, and be pleasing to the ears, the sound adds nothing to the heat. Similarly, a foolish person rejects the Lord for earthly pleasure and laughter, both of which are ultimately meaningless. They aren't sustained.

Skin on fingers and arms is often pricked by plant thorns when the thorny burnet is added to a fire. Similarly, when foolish individuals laugh at us—even when the laughter is short lived—the laughter pricks us. True, we can look back and reflect on whether or not their loud laughter was appropriate or valuable; but, in the moment, the laughter hurts.

****

Ridicule and laughter can hurt us.

1. People who make fun of others act foolishly. What does the fool say about God (Psalm 14:1)?

2. Is God with you when you are the butt of foolish laughter? Where is God at that time (Psalm 14:5)?

3. If you wanted God to be with you – so you can ignore when others make fun of you—what can you do?

****

Outcome

Ecclesiastes 7:6 is the only Bible verse which identified the thorny burnet; however, since the days of Adam and Eve, thorns and
thornbushes have played a role in agriculture and in parables. In the Bible, thorny plants were often associated with desolation and ruin. Enjoyment such as alcohol and carousing with friends leads to foolish laughter; but it also leads to desolation and ruin.

Sometimes life on earth seems long, as year after year goes by. Yet, these seemingly long years are short when compared to eternity. So, too is the seeming happiness of foolish decisions. For a short time, they may bring gaiety; but, in the long run, which includes eternity, they provide little joy.

Takeaway

I feel inadequate, even intimidated, when individuals make fun of me and my faith. An alternative response is to pray for them. At the same time, I can feel superior to them.

Chapter 7

THE CEDAR & THISTLE

Getting Started: Read 2 Chronicles 25:17-18

The parable of the cedar and thistle is technically a fable. In it, one plant speaks to another, i.e., the thistle to the cedar of Lebanon. Some Bible commentators called this parable a riddle because certain words King Jehoash of Israel (Northern Kingdom) spoke were difficult to interpret. When King Jehoash delivered the parable to King Amaziah of Judah (Southern Kingdom), he gave a warning to Judah's king.

Backstory

Preceding history includes events in both Israel and Judah. This "Backstory" is long because it involves happenings in two kingdoms.

Initially, King Amaziah followed God. He determined to finish subjugating Edom (mostly composed of Moab), started by his father King Joash. King Amaziah mustered an army of 300,000 fighting men from Judah and Benjamin. In addition, he hired 100,000 soldiers from Israel, paying them upfront a hundred talents of silver.

Then, a prophet from God came to King Amaziah. The prophet told King Amaziah that if he took Northern Kingdom soldiers with him into battle, God would cause him to lose the battle with Edom. King Amaziah believed the prophet. He sent the Israelite soldiers home, allowing them keep the money he paid them.

Northern Kingdom soldiers were furious. They wanted the opportunity to fight in the battle and to plunder rich Edomite lands. Plunder, not payment for fighting, was the major source of riches for soldiers in the ancient near east. While King Amaziah was fighting the Edomites, these same Northern Kingdom soldiers raided towns that belonged to Judah. They killed three thousand individuals and carried off great quantities of plunder.

Meanwhile, King Amaziah led his army to the Valley of Salt where they killed 10,000 Edomites outright. They captured and killed another 10,000 Edomite soldiers.. When King Amaziah returned from battling the Edomites, he brought back Edomite idols. King Amaziah set the idols up in Jerusalem, bowed down to them, and burned sacrifices to them.

God was angry with King Amaziah for his idol worship and sent a prophet to him. The prophet asked the king, "Why do you consult this people's gods, which could not save their own people from your hand?" (2 Chronicles 25:15). While the prophet was speaking, King Amaziah interrupted him and told the prophet that if he continued to speak he would be killed. The prophet stopped but told King Amaziah that God would destroy the king because he worshiped idols and wouldn't listen to the prophet's counsel.

****

Idolatry is always wrong.

1. Why did King Amaziah bring the Edomite gods/idols back to Jerusalem and worship them?

2. Why consult a god of a conquered people?

3. Do you have any gods/idols that you carry around from place to place that replace God?

****

After consulting with his advisors, King Amaziah sent a message to King Jehoash (king of the Northern Kingdom) to meet him face-to-face. In other words, Amaziah challenged Jehoash to battle. Likely, the reason for Amaziah's challenge was Israel's plunder of Judah's lands while Amaziah was fighting the Edomites. Because a king's subjects paid taxes, they expected their king to keep them safe. As king of Judah, Amaziah couldn't ignore Israel's murder of 3000 individuals or their plunder of Judah's land.

Parable and Warning

Rather than immediately agreeing to Amaziah's challenge, King Jehoash sent a message back to Amaziah in the form of a parable and a warning:

> A thistle in Lebanon sent a message to a cedar in Lebanon, Give your daughter to my son in marriage. Then a wild beast in Lebanon came along and trampled the thistle underfoot. You say to yourself that you have defeated Edom, and now you are arrogant and proud. But stay at home! Why ask for trouble and cause your own downfall and that of Judah also? (2 Chronicles 25:18-19).

In the parable, King Jehoash of Israel referred to Israel and himself as a mighty cedar of Lebanon. In contrast, he named King Amaziah and Judah as a lowly thistle. By identifying himself and Israel as cedars of Lebanon, King Jehoash co-opted for Israel the name "cedar of Lebanon," a term used to refer to the Israelite nation from before they entered the Promised Land.

For God's temple, Solomon used cedar trees from Lebanon rather than any tree in his own country, i.e., sycamore or box tree.

Characteristics of Cedars and Thistles

In the Bible, about 20 different words are related to some type of prickly or thorny plant. In Jehoash's parable, the Hebrew word for thistle is *choâch* or *hoah* and is associated with the *Scolymus* genus of plants. Given different characteristics and value of the two plants, who would want to be characterized as a thistle when the alternative was the cedar of Lebanon?

When King Jehoash named King Amaziah a thistle, possibly he was thinking of the spotted golden thistle. There is a huge difference in size and characteristics of this thistle and cedar tree. In contrast to the spotted golden thistle which grows a maximum of 3 feet tall, the cedar can grow 120 feet and has a main trunk diameter of six-to-eight feet. The spotted golden thistle is an annual herbaceous plant that grows in abandoned fields and ditches and along paths and trails. Essentially, thistles are weeds. In contrast, cedars are evergreen trees that are beautiful in winter and summer. The cedar of Lebanon grows in mountainous and rocky soil; its roots extend deep into the ground. Cedars grow slowly; it takes centuries to produce a majestic cedar tree.

Spotted golden thistles have little if any value to man or animals. Leaves are oblong and prickly. For the most part, even livestock avoid the thistle. Yellow flowers bloom May to August but have no noticeable scent. Cedar wood is durable, can be free from knots, and easy to work. The heart wood is warm red and beautifully grained. Cedars exude gum or balsam which gives the tree an aromatic scent in which people take delight. Most insects dislike the smell and taste; consequently, they don't attack cedars. Cedar wood is resistant to fungal disease so doesn't readily develop dry and wet rot.

To King Jehoash, King Amaziah's challenge was as ludicrous as if he asked for Jehoash's daughter to wed his son. In ancient times, sons of kings married daughters of kings only if the kings were equals in terms of land and wealth. Clearly, King Jehoash perceived that there was no equality between himself and the King of Judah. He was the mighty cedar of Lebanon and a much more powerful king than Amaziah of Judah.

King Jehoash went further in his insult to King Amaziah by giving King Amaziah a warning: King Jehoash identified King Amaziah as arrogant because he defeated the Edomites; but King Amaziah was asking for trouble if he persisted with the challenge to Israel. A future battle would result in King Amaziah and Judah's downfall.

King Jehoash was convinced that even though King Amaziah won a decisive victory over the Edomites, he and his army were more experienced. After all, they won three major battles against Aram, a noteworthy adversary. Further, Judah and Benjamin were composed of only two tribes versus the ten tribes of Israel. Israel had more fighting men than Judah.

The spiritual interpretation of the parable King Jehoash offered to King Amaziah was that King Amaziah's pride would lead to his destruction. King Amaziah didn't comprehend that his success over Edom was due to God being on his side. After King Amaziah's decision to worship Edom idols and not allow God's prophet to advise him, God left King Amaziah. Now, he would go into battle against King Jehoash and Israel with no divine assistance.

Cedar of Lebanon
Golden thistle
Maria Lin

****

Silently, we glory in our accomplishment.

1. Has your high self-assessment ever caused you to think more highly of yourself in a church situation?

2. Has your low self-assessment ever caused you to think more highly of yourself in a church situation?

3. How can we have a realistic assessment of our abilities?

****

Outcome

The two kings met in battle at Beth Shemesh, a town in Judah near the border of Dan. The Bible gave no details about the battle, i.e., number of men on either side; however, King Jehoash captured King Amaziah. King Jehoash took King Amaziah to Jerusalem. There, King Jehoash tore down 600 feet of Jerusalem's wall in an area that made Jerusalem most vulnerable to attack. Jehoash took gold and silver from God's temple along with the palace treasures. King Jehoash returned to Samaria with hostages from Judah. Some scholars believe King Jehoash took King Amaziah with him to Samaria. Only after King Jehoash died was King Amaziah returned to Judah.

King Amaziah lived about fifteen years after King Jehoash died. From the time Amaziah turned away from God, Jerusalem leaders conspired against him. At some point, King Amaziah fled Jerusalem, fearing for his life. Jerusalem conspirators sent men to Lachish and assassinated King Amaziah. King Amaziah's life and rule became as insignificant as a thistle which King Jehoash compared him to in the parable.

Takeaway

When pride comes, then comes disgrace, but with humility comes wisdom (Proverbs 11:2).

Chapter 8

POEM OF THE PLOWMAN

Getting Started: Read Isaiah 28:24-29

The parable of the plowman is a two-stanza poem. The first stanza focuses on plowing the ground for planting; second stanza on threshing a crop. The poem ends with praise to God who taught humankind how to plant and harvest. Finally, the parable's interpretation unfolds.

Because the Hebrew language and Hebrew poetry differs from English, native English-speakers can't appreciate the memory cues in this Hebrew poem. We don't remember the poem in the same way that we remember, even memorize, Burns, Longfellow, and Whitman's poems.

The parable of the plowman resonated with farmers. The writer, Isaiah, was an urban agriculturist. Although Isaiah lived most of his life in Jerusalem, the content of Isaiah's book demonstrates a sure knowledge of ways crops were planted, cultivated, and harvested.

Isaiah saw visions and prophesized during the reigns of Kings Uzziah, Jotham, Ahaz, and Hezekiah. From Jerusalem he saw the final days of the Northern Kingdom. He warned the Southern Kingdom (Judah) that her sins would bring captivity. At the same time, Isaiah assured hearers that a remnant of Jews would be restored to Jerusalem and Judah.

Backstory

The background for the parable of the plowman begins in Isaiah chapter 28, verse 9. The country's leaders mocked Isaiah. More importantly, they mocked God! The leaders asked: Who is God to try to teach us? Are we babies? Are we newly weaned children? No! They were adults and didn't need detailed instructions from God, i.e., "Do and do, do and do, rule on rule, rule on rule, a little here, a little there" (Isaiah 28:10). The rulers in Jerusalem boasted they entered a covenant with death and the grave; thus, when a foreign nation invaded them, they wouldn't be touched.

****

Most Christians believe God speaks to them.

1. Many years ago, I was praying and believe strongly that God spoke to me in my prayer. When I shared my experience with a non-Christian friend, her comment was ridicule, i.e., there was something mentally wrong with me because I heard voices. What do you think?

2. Does God speak to you? What are ways of God speaking to you?

3. What do your mediation/contemplation/prayers consist of? Do you ever take time for God to speak to you as well as speaking to God?

****

According to Isaiah, reality for both Israel and Judah would be different than what they expected. Neither country was prepared militarily or spiritually for foreign invasion, an invasion that

God allowed in judgment for their rejection of God and his laws. Foreign soldiers would engulf Israel and Judah. Instead of God's words through a prophet, citizens would hear a foreign language spoken on foreign lips. Protection from death and the grave, it there was such a thing, would be annulled. The people will be beaten down by the invaders, taken captive, and exiled. Power politics, i.e., reaching out to Egypt for protection from northern invaders, won't work.

Parable in a Poem

Isaiah implored the Israelites to hear him four times. Isaiah wasn't speaking to hear his own voice; nor was he speaking his own thoughts and words. Instead, Isaiah gave God's message, a message vital to Judah's health and security:

> Listen and hear my voice; pay attention and hear what I say. When a farmer plows for planting, does he plow continually? Does he keep on breaking up work the soil? When he has leveled the surface, does he not sow caraway and scatter cumin? Does he not plant heat in its place, barley in its plot, and spelt in its field? His God instructs him and teaches him the right way.
>
> Caraway isn't threshed with a sledge (sleigh), nor is the wheel of a cart rolled over cumin. Caraway is beaten out with a rod, and cumin with a stick. Grain must be ground to make bread; so one does not go on threshing it forever. The wheels of a threshing cart may be rolled over it, but one doesn't use horses to grind grain. All this knowledge comes from the Lord Almighty, whose plan is wonderful, whose wisdom is magnificent (Isaiah 28:24-29).

****

Who taught/teaches humankind?

1. Do you believe—really believe—that God taught farmers how to prepare soil, plant seeds? Or did farmers learn by a trial-and-error evolutionary process?

2. What are the implication of believing that God taught farmers the process versus trial and error?

3. Do you think that in the 21st century, God teaches mankind new technologies or is mankind's current learning processes trial and error?

****

The parable of the plowman identified three activities a farmer used:

1. Breaking soil
2. Sowing seed
3. Threshing the crop.

The order of these three actions comes from God; they couldn't be implemented in a different sequence and still achieve an optimal harvest. Further, each class of seed requires its own soil, sowing method, and threshing.

Breaking the soil. A farmer plowed his land; but, at a point when the soil was soft and ready to receive the seed, the farmer stopped working the ground. Too fine plowing and repeated plowing wastes the farmer's time and contributes to top-soil erosion and loss. Over working the ground contributed to the Dust Bowl in the 1930s in central United States.

Sowing seed. A farmer knew the best sowing techniques to get an optimal crop. He sowed caraway and scattered cumin. The knowledgeable farmer even knew that each hard grain (wheat, barley, spelt) seeds needed to be sowed differently. He didn't mix seeds together. He planted grain seed in their own plots aware that they matured at different rates and were harvested at different times.

Threshing crops. The farmer knew how to thresh his crops to obtain a maximum yield. He beat the caraway with a rod (flail); however, cumin was extracted with a stick. Wheat, barley, and spelt were threshed using sledges or threshing wheels rotated by a donkey or ox.

Characteristics of Cumin and Barley

In the parable of the plowman, two herbs were named. Although New International Version Bibles translate the first herb as caraway, other Bibles (AMP, ESV, KJV) translate this herb as fitches, dill, or fennel. Because the second herb, cumin, was translated uniformly in different Bibles, I use cumin as the prototype herb. Barley is contrasted with cumin because of barley's historic significance to the Israelite nation.

Although cumin is an herb, it has the distinction of being the second most popular spice in the world. Probably, cumin originated in the Nile River Valley. Israelites may have taken cumin seeds with them when they left Egypt. According to Isaiah, ancient Israelites scattered cumin on top of prepared soil in the spring.

Cumin seeds need a lot of water, but not a soggy environment. Cumin plants benefit from a little crowding. Ideally, they are grown in groups or clumps. Plants support each other, keeping stems from bending over and touching the ground.

Cumin seeds are ready to harvest when pods turn brown and dry. When this happens, pods open and spill their seeds onto the ground. Often, pods don't dry all at once but need to be harvested at different times throughout the season.

In contrast to the herb cumin, barley is a grain. For Israelites, barley was a major food staple for several reasons. First, barley was less expensive to purchase than wheat. Second, barley was drought-resistant and grew in both alkaline and saline soils; consequently, barley grew in the diverse habitats of the Promised Land. Third, under favorable conditions, barley ripened in as few as 60–70 days, a shorter time than wheat matured. In Israel, winter barley was sown in November and December and harvested in April or early May. Israelites were so closely associated with barley that Midianites referred to them as "barley cakes" (Judges 7:13).

By the time Isaiah prophesized, Israelites used two-hand wooden plows with an iron point on the end; the iron point turned over the ground. Plows were pulled by an ox and guided by the farmer. Barley seeds

were dropped in rows or furrows and covered by harrowing (or smoothing) the soil over seeds. Once ripe, barley stalks were cut using a scythe; then transported to the threshing floor in wagons or in large baskets.

Wealthy farmers had their own threshing floors for grains, but more often a threshing floor was shared by a village or town, i.e., they were publicly owned and located outside of the town wall.

When Isaiah prophesized, there were four methods of threshing, i.e., stick, animals, threshing sledge, and wheel. The most common and effective method to thresh barley was with the sledge (sleigh). A sledge was a platform made from two wooden boards. The front was upturned so it wouldn't be obstructed by the floor when pulled by animals. On the sledge bottom, hard stones and later iron was embedded to break the barley kernel from stalks.

Cumin
Barley
Maria Lin

Meaning of Plowman Parable

In order to achieve security for their land, citizens had to recognize and abide by God's rules just as planting herbs and grains followed certain agrarian rules. Unfortunately, in Israel and Judah both leaders and citizens rejected God and God's requirements for national and personal holiness. There was little, if any, justice in either Israel or Judah. The result was God's judgement on each nation and its people.

While the wealthy Israelites made bread from wheat, poorer families used barley almost exclusively. For the poor, wheat flour was a luxury. As both the Northern and Southern Kingdoms aged, the separation between the wealthy who ate wheat bread and the poor who made bread from barley grew. Kings and leaders, insulated by their wealth, had almost no understanding of the everyday lives of the poor in their nations.

****

Ponder national vs individual sins.

1. Differentiate between national and personal sins. Can your nation sin and an individual not sin? Can an individual sin and her nation not you sin?

2. When we confess our sins, do we normally confess our individual sins or national sins?

3. How can we confess sins of our nation to God? What can we as individuals do about national sins?

****

The analogy for the farmer selecting varied methods to sow different types of seed, i.e., herbs versus grains, demonstrates that God's judgements aren't arbitrary. God doesn't stereotype his people. Carefully, he selects the judgement and discipline suited for a person's needs. God uses the lightest possible touch. He never allows his judgment to be greater than his people can bear. Israel's sins were so detestable that the nation was destroyed. In contrast God punished Judah for sins, but didn't punish them forever. The punishment had an endpoint. A glorious future awaited the redeemed nation. As a farmer doesn't plow a field forever, so Judah's punishment won't last forever.

****

God doesn't have a set punishment for sin. God punishes each individual who sins according to his/her level of knowledge and abilities. Is that fair (just)?

****

Outcome

Often Bible commentators compared Isaiah's parable of the plowman to Jesus' teachings in the New Testament. Several asserted that Jesus'
Parable of the wheat and tares echoed Isaiah's plowman parable. Others suggested that Jesus' teaching in Matthew 16:2-4 where the Pharisees and Sadducees couldn't interpret signs of the times had its origins in Isaiah's plowman parable. It makes sense that Jesus based his teachings and parable on Old Testament writings because the Holy Spirit—the third entity of the Trinity—gave the parables to Old Testament prophets and writers.

Growing conditions for cumin seeds parallel how Israelites—particularly the Northern Kingdom—over-indulged themselves. Just as some water is good for cumin seeds and seedlings, some wealth and leisure are good for a nation, but excesses lead to sin. Similar to a cumin plant needing the support of other plants to grow optimally, Northern and Southern Kingdoms, i.e., all 12 tribes of Israel, needed each other for optimal growth. When Israel and Judah separated they didn't have each other for support. Both nations became vulnerable to outside attack. Like cumin seeds, the sins of the Northern and Southern Kingdom matured at different rates. The Northern Kingdom was destroyed almost 200 years before Jerusalem.

Takeaway

Scripture tells us that we reap what we sow. Perhaps we also reap how we sow.

Chapter 9

BASKETS OF FIGS

Getting Started: Read Jeremiah chapter 24

Jeremiah's parable of two baskets of figs contrasted Judah captives taken to Babylon with those who remained in Jerusalem and Judah. The captives taken to Babylon were named "good;" while those who remained in Judah were identified as "bad." Rarely, in history are captives designated the "good guys;" but, if we have learned anything about God, it's that he turns our preconceived ideas of right-wrong, rich-poor, and good-bad upside down.

Backstory

By the time of this parable, Jeremiah had been God's prophet for about 23 years. In the years prior to Jeremiah's parable of the two baskets of figs, Judah's King Jehoiakim was murdered. His son, Jehoiachin, succeeded his father to the throne. After ruling three months and ten days, the eighteen-year-old king surrendered when Nebuchadnezzar besieged Jerusalem. This grandson of godly King Josiah, his mother and wives, capable fighting men, and the most skilled artisans and craftsmen were taken captive (597 BC) to Babylon. Nebuchadnezzar made Jehoiachin's uncle, Zedekiah, the vassal king in Judah.

Parable in a Vision

In the parable of the two baskets of figs, God gave Jeremiah a vision that included a parable. The reader isn't sure if Jeremiah was at the front of the Jerusalem temple when he saw the two baskets of figs, or if he was elsewhere and God included both the temple and baskets of figs in the vision. Here's how the Bible recorded Jeremiah's parable:

> After Jehoiachin son of Jehoiakim king of Judah and the Lord showed me two baskets of figs placed in front of the temple of the Lord. One basket had very good figs, like those that ripen early; the other basket had very bad figs, so bad they could not be eaten (Jeremiah 24: 1-3)

Jeremiah noticed that one basket contained good figs, they looked succulent and edible. In contrast, the second basket was bad. The Bible didn't specify how the second basket of figs looked "bad". Perhaps, the figs were mushy, spoiled, or rotten; or, they could have been small and dry. Whatever was wrong with these figs, they couldn't be eaten:

> Then the Lord asked me, What do you see, Jeremiah? Figs, I answered. The good ones are very good, but the bad ones are so bad they cannot be eaten (Jeremiah 24:3-4).

The question that God asked Jeremiah in this passage is the same one that God asked Jeremiah in an earlier vision of the almond branch: "What do you see?" As with the first vision, God expected his prophet to not just look at, but to understand, the vision. Jeremiah's response was accurate, i.e., he saw a basket of good and a basket of inedible, bad figs.

Then, God told Jeremiah the meaning of the vision of two baskets of figs:

> Then the word of the Lord came to me: This is what the Lord, the God of Israel, says: Like these good figs, I regard as good the exiles from Judah, whom I sent away from this place to the land of the Babylonians. My eyes will watch over them for their good, and I will bring them back to this land. I will build them up and not tear them down; I will plant them and not uproot them. I will give them a heart to know me, that I am the Lord. They will be my people, and I will be their God, for they will return to me with all their heart (Jeremiah 24:4-7).

****

God expected the captive Jews not to resent their captivity or captors.

1. Describe how you would feel if you were uprooted from your home and taken captive into another country.

2. How hard would it be for you to agree with God that you deserved exile as your punishment?

3. How should refugees be treated by Christians in the country where they seek asylum?

4. Most American resonate to New Hampshire's state motto, "Live free or die." Yet, God told the Jewish exiles to do the opposite, i.e., submit to and live under the yoke of Babylonian oppressors.

Characteristics of Fig Tree and Figs

Several characteristics of fig trees and their fruit make them a good illustration for Jeremiah's vision of good and bad baskets of figs. Although a relatively small tree, fig trees figured prominently in Israelite lore and holy writings. Supposedly, Adam and Eve used fig leaves to first clothe their bodies. The fig leaves in the Adam and Eve account symbolized their disobedience to God in much the same way that most of Judah's people disobeyed God's laws.

Fig leaves are large (typically 5-10 inches long and 4-5 inches wide) and attractive, but they are rough and hairy on the underside. When injured, leaves exude a bitter, sticky, and milky resin. Adam and Eve sowed several fig leaves together to create aprons to cover their naked bodies. These fig leaf clothes would have been itchy, sticky, and altogether uncomfortable against the skin. Judah's choices to disobey God and hide in their sin resulted in much more than mere discomfort.

Through Moses, God told the Israelites that the fig tree would be one of the seven species of plants that would enrich their lives in the Promised Land; yet fig trees aren't exceptionally hardy. They grow best in a narrow environment. The average fig tree grows about 20 feet tall and develops a spreading canopy. The trunk is short; bark is gray and smooth. The tree trunk is sensitive to heat and sun and easily damaged. Further, fig trees can be damaged if temperatures drop to 30 degrees Fahrenheit. Perhaps, the bad figs in Jeremiah's parable came from a tree that received too much summer heat or were damaged by early frost.

Sometimes figs grow under leaves and at other times above leaves. In the Middle East, fig trees bear fruit twice a year. The winter (first) fig crop ripens in June; the summer (second) crop ripens August through September. Often on fig trees, fruit (figs) appears prior to, or with, leaves. Remember, during Passover, Jesus walked to a fig tree on his way from Bethany to Jerusalem. Because he saw leaves on the fig tree, Jesus expected the tree would contain figs he could eat. The fig tree contained leaves, but no figs. As with Jeremiah's basket of bad figs, Christ concluded, the fig tree was useless. The proper response was to cut down the useless tree.

Unique among flowering trees, fig flowers aren't seen; rather, flowers are enclosed inside the fig (fruit) itself. Similar to the hidden fig flower, God wanted Judahites to produce good works, not for outward display, but, to sweeten the lives of individuals around them. Unfortunately, the last four kings of Judah evidenced little, if any, good fruit in their lives. Individual citizens of Judah followed the lead of their kings.

After four years of growth, most fig trees begin to produce fruit. Ideally, figs ripen fully before they are picked. A fig is ripe when it darkens, turns brown or purple, starts to bend at the neck, and is slightly soft to the touch. Because all figs don't ripen at the same time, harvest occurs over several months. To get the basket of bad figs, Judahites may have picked all of the figs at the same time. They didn't realize or care that some figs weren't ripe and wouldn't ripen off the tree branch.

Meaning of Parable in Vision

Clearly, God regarded the exiles taken captive to Babylon as good figs. God planned to watch over them during their captivity and bring them back to Judah. In addition, God will give the captives a heart to know him, and they will turn to God with all their heart.

The balance of Chapter 24 is God's interpretation of the basket of bad figs which can't be eaten:

> But like the bad figs, which are so bad they cannot be eaten, says the Lord, so will I deal with Zedekiah king of Judah, his officials and the survivors from Jerusalem, whether they remain in this land or live in Egypt. I will make them abhorrent and an offense to all the kingdoms of the earth, a reproach and a byword, a curse and an object of ridicule, wherever I banish them. I

Good Figs
Bad Figs
Maria Lin

> will send the sword, famine and plague against them until they are destroyed from the land I gave to them and their ancestors (Jeremiah 24:8-10).

In contrast to the good figs, the outcome for the bad figs was dire. The bad figs were King Zedekiah, his officials, and other survivors of the Babylonian siege of Jerusalem. God was going to send sword, famine, and plague on people who remained in Judah. Indeed, during the siege of Jerusalem, residents suffered famine and pestilence. Soldiers murdered thousands of Jerusalemites. Survivors were exported to foreign kingdoms. God determined to make these exported Judahites abhorrent to the people of every kingdom where the Babylonians settled them.

****

Exile was the price the Jews paid for disobeying God through idol worship and treating fellow Israelites with contempt.

1. Despite their rebellion against God, many Judahites continued to make sacrifices at God's temple. Today, we don't make burnt offerings to God. Instead. we give money to the church and worthy charities. Are we mimicking the behavior of Judahites? Explain your answer.

2. What if most of our actions are godly, but inside we are angry and resentful? Is exhibiting godly behavior sufficient, or do we have to have a change in our heart? How can we change our hearts?

3. Samuel told the Israelites that to obey God is better than to make sacrifices to him? How would you apply Samuel's words to your life?

****

Jeremiah's prophecy of the good and bad figs came true. When Jehoiachin arrived in Babylon, he was placed in prison. There, he remained 37 years. When Nebuchadnezzar died, his son Evil-Murdock became king over Babylon. King Evil-Murdock released Jehoiachin from prison, gave him an allowance, and a favored place at the king's high table for meals. Seventy years after Jehoiachin's captivity, his grandson, Zerubbabel, led 50,000 Jews back to Jerusalem.

Outcome

Meanwhile back in Jerusalem, King Zedekiah rebelled against Babylon. Nebuchadnezzar returned to Jerusalem, laid siege to the city, killed King Zedekiah, and conquered Jerusalem and the surrounding towns. He destroyed Jerusalem's walls. Nebuchadnezzar assigned Gedaliah, a politically moderate Jew, as governor of Judea. Governor Gedaliah established his capital at Mizpah.

Scores of years later, exiled Jews were allowed to return home to Judea and rebuild Jerusalem. This return only occurred after the Persians conquered the Babylonians. When Cyrus ordered the Jews to return home to Jerusalem, he was convinced he acted as God directed.

Takeaway

A number of years ago, I had a big professional disappointment. I was devastated even though I knew I was at fault. In the Northern and Southern Kingdom, individuals must have felt the same way when they were exiled. I cried out, "Just tell me what you want me to learn from this situation so I can get past the mental pain." I wanted my pain to go away more than I wanted to learn God's lesson.

Chapter 10

VISION IN MYRTLE TREES

Getting Started: Read Zechariah 1:8-13

The parable of the man among the myrtle trees was a prophetic vision, designed to appeal to the eyes rather than the ears. Jews regarded the myrtle as a symbol of something good and beautiful; yet, the myrtle tree couldn't compare to the cedar of Lebanon. The parable was set in Judea, an insignificant province of the Persian Empire.

The first year (535-536 BC) after the Jews returned to Jerusalem from Babylonia, they rebuilt the temple altar. The second year, they laid the second temple's foundation. Non-Jewish people who lived in the area, largely Samaritans, offered to help rebuild the temple. When the Jews refused their assistance, these enemies initiated a systematic program to discourage the Jew's efforts. As a result, temple construction slowed, then, stopped for about 10 years.

In August 520 BC, God spoke through the Jewish prophet Haggai. Prophets were God's voice on earth. They spoke or recorded exactly what God said, rather than paraphrasing or interpreting God's message. God told the Jews to finish building the temple. Haggai identified that the drought in Judea was the result of God's temple remaining in ruins. Although returned

Jews built their own homes, they didn't rebuild God's home. Almost immediately, Jews reinitiated temple construction.

Two months after God's message to Haggai, Zechariah started to receive messages from God. Zechariah's prophecies mirrored Haggai's, i.e., rebuild the temple; but, included that the Jews repent and serve the Lord. Zechariah was God's prophet for the next 40 years (520-480 BC).

Parable in a Vision

In the first year of his ministry, Zechariah received eight visions from God in one night. They were visions, not dreams; Zechariah was fully awake. The first vision was a parable that included a plant, the myrtle tree. Here is Zechariah first vision:

> There before me was a man mounted on a red horse. He was standing among the myrtle trees in a ravine. Behind him were red, brown and white horses. I asked, "What are these, my lord?"
>
> The angel who was talking with me answered, 'I will show you what they are.'
>
> Then the man standing among the myrtle trees explained, 'They are the ones the Lord has sent to go throughout the earth. And they reported to the angel of the Lord who was standing among the myrtle trees. We have gone throughout the earth and found the whole world at rest and in peace.'
>
> Then the angel of the Lord said, "Lord Almighty, how long will you withhold mercy from Jerusalem and from the towns of Judah, which you have been angry with these seventy years?"

So the Lord spoke kind and comforting words to the angel who talked with me.

Then the angel who was speaking to me said, 'Proclaim this word: This is what the Lord Almighty says: I am very jealous for Jerusalem and Zion, and I am very angry with the nations that feel secure. I was only a little angry, but they went too far with the punishment. Therefore this is what the Lord says: I will return to Jerusalem with mercy, and there my house will be rebuilt. And the measuring line will be stretched out over Jerusalem, declares the Lord Almighty. Proclaim further: This is what the Lord Almighty says: My towns will again overflow with prosperity, and the Lord will again comfort Zion and choose Jerusalem' (Zechariah 1:8-1).

****

If you were a postexile Jew living in Jerusalem, how would you respond to Zechariah's parable of the man among the myrtle trees.

1. Would you have been encouraged or discouraged by Zechariah's parable? Explain your reasoning.

2. Would there have been any part of you that wondered how God could allow Babylon to enact such vicious brutality on the people of Jerusalem? Possibly, Jerusalemites that lived through the siege and the sacking of Jerusalem saw beloved parents or grandparents die or be murdered.

3. Would you have been a) angry with God, b) grateful that God cared about you and your people enough to discipline you, or c) totally confused about God's supposed goodness?

****

Characteristics of Myrtle Tree Growth

At the time of Zechariah's prophecy, myrtle trees grew wild throughout Judea. The myrtle is an evergreen shrub or small tree, i.e., under optimal conditions myrtle trees grow up to 24 feet. The myrtle tree grew under the canopy of taller trees or where taller trees were harvested or destroyed. Using the myrtle tree to symbolize Jews, represented that both in the present and in the future, Jews were a small nation. No longer was the Jewish nation comparable to the 120-foot cedar of Lebanon.

The myrtle leaf is dark green, oval, and pointed. Leaves are fragrant because the leaf tissue contains oil-secreting glands. Myrtle leaves and flowers have attractive scents which are often described as sweet, flower-like, aromatic, and unique. Like a leaf from the insignificant myrtle tree, Jews could still give off a sweet odor pleasing to God.

In Zechariah's parable, myrtle trees grew in a ravine. A ravine is a narrow steep-sided valley, a low place. Myrtle trees don't tolerate shade; yet, shade is a common characteristic of a ravine or gorge. Further, myrtle trees are damaged by cold dry wind, when young they can be damaged by frost. In a ravine—the microclimate and setting of Zechariah's parable—young myrtle trees could be damaged by shade, cold, and wind blowing through the ravine. Similarly, the Jewish remnant in Judea could be easily damaged, even destroyed, by conquerors. Future Jewish nations could thrive, but only in the shadow of mighty nations.

The Hebrew word for myrtle tree is *hădaç*. Little is known about the derivation of *hădaç*; however, it may mean "to grow rapidly." In contrast, the word for cedar is *'erez* meaning "to be firm." Firm means securely or solidly fixed in place and having a structure that resists pressure. The cedar tree was firm. Its tenacious root structure was tenacious, it lived long, and was resistant to insect infections.

Myrtle
Maria Lin

****

Advantages of being a myrtle tree??????

Identify some advantages of being a myrtle rather than a cedar tree.

1.

2.

3.

Meaning of the Parable in the Vision

Like the fast-growing myrtle tree, the population in postexile Judea grew rapidly. Some population increase was due to birth rates; but, more probably, Jewish population increased because Jews moved into Judea from other parts of the Persian Empire.

With God's assistance the Jews grew again into a nation; however, like the myrtle tree, the Jews didn't have the national durability and firmness seen in King David and Solomon's reigns. The once mighty cedar of Lebanon was gone; it was now a somewhat lowly myrtle tree. Judea was a small province in the outback of the expansive Persian Empire. Going forward, most of Judean history showed that the country was controlled by foreign nations, i.e., Rome, Turkey, Great Britain.

Despite the diminished importance of the Jewish kingdom, Zechariah's parable was a message of hope and encouragement to postexile Jews. The interpretation was that God keeps his promises. God comforted the angel who was distressed about the state of the lives of the postexile Jews. God promised that the Babylonian-Persian Empire would be held accountable for their brutal treatment of his people when Judah and Jerusalem were conquered. Further, God promised that he would return to Jerusalem and Zion with mercy and the temple would be rebuilt.

****

God promised that Jews living in the tiny Judean province would prosper.

1. The Bible revealed some reasons why people prosper and reasons why they don't prosper in the following verses. Read and summarize reasons for prosperity from these Bible verses:

    a. Deuteronomy 28:62-63

    b. Proverbs 17:20

    c. Proverbs 28:13

2. How can we meld our personal experiences with what the Bible says, yes even promises, about prosperity being related to a godly life?

3. According to God's word, why have you prospered or not prospered?

****

Outcome

The Persian Emperor Darius III died about 330 BC, leaving Alexander of Greece ruler of much of the remaining lands of the once mighty Persian Empire. About 200 years had passed between God consoling the angel of the Lord and God holding the Babylonian-Persian Empire accountable for their brutal treatment of the Jerusalem Jews. Within those 200 years Persia fought and lost many battles with vicious enemies where soldiers and civilians were ruthlessly killed. We are reminded anew that God keeps his promises, but God doesn't operate within the same time constraints as man.

Takeaway

Does God need to return to the United States of America in a way similar to his promise to return to Jerusalem? If so, what is your role in making this happen?

Section 2

PLANTS IN NEW TESTAMENT PARABLES

In contrast to the few books written on Old Testament parables, there is an overabundance on New Testament parables. Almost all parable books focus on Jesus' parables; however, other New Testament figures told parables, i.e., John the Baptist, Paul, James, Peter, and John the Apostle. Many of these writers used plants to show the meaning of the of their spiritual message. Section 2, Plants in New Testament Parables, contains 10 parables from the New Testament which were illustrated by plants. Six parables were spoken by Jesus and four by other New Testament figures.

The New Testament was written in Greek; the Greek word *parabŏlē* was used for parable. *Parabŏlē* translates as a comparison, placing beside, and fictitious narrative usually of common life conveying moral truth. Some New Testament truths were made using plants in these simple stories. Others were acted out or seen in visions.

Parable development reached its zenith in Jesus' parables; about 1/3 of his words were parables. But, Jesus didn't start his ministry telling parables. Think back to Jesus's Sermon on the Mount. Despite covering several chapters, the Sermon contains few parables. Imagine the disciples' surprise when one day Jesus' teaching method changed. He started to teach almost exclusively in parables. The disciples were as confused as the crowds who came to hear Jesus. They asked Jesus why he was using parables.

Jesus used parables for three reasons. First, parables prevented the Jewish leaders—many who were or would become Jesus' enemies—from using his own words to condemn him. Second, Jesus use of parables fulfilled Isaiah's prophecy that people wouldn't understand, when even the Messiah spoke, because they were spiritually dull. Third, parables contained truths that could only be understood through God's revelation. Although God wanted all individuals to understand the parables, the sad truth was some wouldn't understand because they weren't willing to accept Jesus and his message. But, to those individuals who were willing to accept Jesus as Messiah, interpretation of parables provided insight into the kingdom of God.

Jesus' teachings never seem outdated. His parables are as relevant as "alerts" that flash on FOX or NBC news networks. Jesus spoke seven parables in Section 2, but they aren't an exhaustive list of Jesus' parables. They are included because they were the foundation of what Jesus wanted people to know about the kingdom of heaven and because they don't repeat any plant already described in Section 1, Plants in Old Testament Parables.

Jesus and John the Baptist lived and taught in Galilee, Judea, and in the trans-Jordan area known as Perea. Collectively, these areas are called the Holy Lands. Logically, both thought of plants in their own country when they included plants in parables. Other New Testament parable givers, i.e., Paul, the author of Hebrews, Peter, and John (who authored Revelations), probably wrote from other parts of Asia Minor or southeastern Europe. Nevertheless, Hebrew scholars and botanists consistently identified Holy Land plants when naming plants used to illustrate these parables.

Chapter 11

WINNOWING WHEAT

Getting Started: Read Matthew 3:12

In the early part of the first century, residents of Judea and Galilee were fairly prosperous. No wars ravaged these regions which were part of the Roman Empire. Most residents knew how to work within the Roman political and economic systems. The sentiment "just keep things level," reflected the self-interest and expediency of residents.

****

Lessons from hummingbirds

Bruce and I stayed for a long weekend at a small resort in Canada named Destiny Bay. Over long dinners, we watched hummingbirds frenetically fly from one feeder to another, drinking the nectar.

1. The hummingbirds were so busy that they never slowed down to think about their destiny. How is their behavior a reflection of yours?

2. Think about standing before the judgement seat of God. What do you think will happen there?

3. What are you going to say when God asks you to give an account of your behavior here on earth?

****

Backstory

A prophet, John the Baptist, was born to Zechariah and Elizabeth. His hometown was in the hill country of Judea. Zechariah was from the family of Abijah, a division of priests that rotated to Jerusalem to burn daily incense in the temple. Realistically, John followed in Zechariah's footsteps and became the village priest. As a member of Abijah's priestly family, John probably had at least one rotation in the Jerusalem temple prior to his sojourn in the wilderness.

John preached personal acknowledgement and repentance of sins followed by baptism—full body emersion—in water as an outward sign of repentance. The baptismal water symbolically washed sins away. Because of John's actions, some Jews recalled the words of the prophets about a coming Messiah. Many yearned for the Messiah's arrival.

When these people heard about John and saw his actions, they asked John if he was the predicted savior. John answered their question with firm denial saying that the coming Messiah was mightier than he. John wasn't worthy to take off or carry the sandals of the Messiah. John went on to say that the Messiah will baptize people with the Holy Spirit and with fire.

John didn't stop exhorting the Jews to repent and be baptized. John exhorted those baptized to change their behavior and bear fruit consistent with repentance. When those baptized asked him what they should do, John's answer wasn't that they quit their jobs and preach God. Rather, in their lives and jobs they should act honorably, treat others fairly, and share with the less fortunate. For example, John told men with two tunics to give one to the man who had none. Soldiers should stop accusing people falsely and extorting money from them.

John the Baptist spoke bluntly to the multitudes that came to him for baptism; but reserved his worst denunciations for Jerusalem's elite, i.e., Pharisees, Sadducees, and priests. He called them "vipers". The fierceness of his words suggests firsthand knowledge of their behavior. Perhaps, when he rotated through the Jerusalem temple as a priest, John saw the excesses, insincerity, and, yes, even corruption that infected priests of the Herodian-style Jerusalem temple.

****

Is anyone above the need to repent?

1. Do you think the elite, upper-crust Pharisees, Sadducees, and wealthy Jerusalemites in first century Palestine believed that a country priest/prophet's words applied to them? Why or why not?

2. Why did John label these men as "vipers?"

3. How would you compare most of today's intelligent and talented politicians and pundits to the Pharisees, Sadducees, and Jerusalem leaders who came to see John?

****

John's Parable

John the Baptist spoke several parables that used plants to make the comparison between the physical reality and spiritual truth, i.e., "the axe is laid to the root of the tree. Every tree therefore that does not bear good fruit is cut down and will be thrown into the fire" (Luke 3:9). His parable on winnowing described separating chaff from wheat kernels:

> His winnowing fork is in His hand, and he will clear out and clean his threshing floor, gathering his wheat into the barn, and burning up the chaff with unquenchable fire (Matthew 3:12).

Characteristics of Winnowing Wheat

In ancient Judea, wheat kernels (seed, grain) were separated from the chaff (stalks, straw) on threshing floors. Generally, chaff was unusable except as fodder for livestock. The farmer separated wheat kernels from chaff using a process called winnowing. Winnowing consisted of throwing the threshed material (chaff and grain) into the air with a fork or a flat basket. The wind separated the valuable grains of wheat from the chaff. Because wheat kernels were heavier than chaff, they fell to the ground or back into the basket. The lighter chaff, dirt, etc., were blown away by the wind. At times, farmers used fans to create air currents to blow chaff and other impurities away from valuable wheat kernels.

By the New Testament era, wheat was used for bread, other baking, and cereal. Ancient peoples called wheat the "giving grain" and associated it with the cycle of birth, life, and death. For the ancient Hebrews, wheat symbolized life from God, primarily physical life. For New Testament era people, John could have identified no better plant than wheat to contrast righteous and unrighteous individuals.

From earliest times, man planted wheat because of its high nutritional—life giving—content. Like barley, wheat was an annual crop; however, wheat required a longer growing season (about 100 days) than barley. Usually, wheat was harvested in May. The Presence Bread for the Tabernacle and Temple was made with fine white wheat flour. God had every right to expect that Jew's actions, especially those of Jewish leaders, were pure similar to the purist wheat flour used in Tabernacle bread.

John the Baptist spoke his parable on winnowing wheat soon after starting his ministry. Yet, John's parable demonstrated a sure knowledge of a different outcome for the righteous versus the unrighteous. Perhaps, John the Baptist's words on winnowing, wheat, and chaff had their origins in Psalm 1:4-5, i.e., the wicked are like chaff that the wind blows away. They won't be able to stand when God judges them. Sinners will be separated from the assembly of the righteous as chaff is separated from wheat kernels.

****

When Christ is done, the threshing floor will be empty.

1. Given John's parable, what is going to happen to men and women who are on the fence, who haven't decided to follow Jesus?

2. Is there going to be an opportunity for a "do over" when Jesus comes the second time? Explain your position using verses from the Bible.

3. What will be your destiny when Jesus separates wheat from chaff?

****

Meaning of John's Parable

When John the Baptist told the parable of winnowing wheat and chaff, the spiritual reality was that the coming Messiah (Christ) would distinguish or separate the righteous from the unrighteous. Jesus would critically analyze people's heart, not pious actions. John said that sincere, righteous individuals—wheat kernels—will be taken and stored in the farmer's barn, i.e., heaven. In contrast, chaff, those with pretend piety, will be forever burned.

Winnowing Wheat
Maria Lin

Knowing that it takes longer for wheat than barley to mature could have reassured John's listeners. John acknowledged that repentance, maturation, and amendment of life take time just as wheat took time to grow.

Outcome

Messiah (Jesus) came to Earth and was baptized by John. John was imprisoned and killed by King Herod. John's body was dead. Physically, John didn't return to Earth and walked on it.

Jesus, the Messiah, reinforced John's parable of winnowing when he told the parable of weeds and wheat being separated at harvest.

Takeaway

God isn't going to accept any chaff into his storehouse. Where does that leave you?

Chapter 12

HARVESTING WEEDS (DARNEL)

Getting Started: Read Matthew 13:24-30, 36-43

The parable of weeds (darnel) in a wheat field points toward the perplexing presence of evil on Earth by using a series of contrasts, i.e., the good farmer-enemy, wheat-tares, and two harvests. The parable was relevant to first century farmers because it described a common occurrence, i.e., wheat and weeds growing in the same field. If a farmer attempted to remove weeds, he likely would destroy the growing wheat. Simultaneously, the parable was prophetic. It describes what will happen at God's final harvest.

****

Many parables focus on a final outcome

1. Why all these Bible writings, even warnings, allocated to

a final outcome for individuals? What so important about the final outcome?

2. Do you ever look at celebrities—politicians, actors, or pundits—think that they are attractive and articulate? Do you envy them? Do you wonder what their relationship is with God? I've come to believe that I need to pray for, not envy, many of these celebrities.

3. Is silence consent? Are we silent about certain social issues because our lives haven't been fertilized enough?

****

Backstory

Matthew is the only gospel writer who included Jesus' parable of the weeds. Because Matthew wrote for a primary Jewish audience, Jews were meant to interpret and apply the parable's content to their lives. Today, Christians can use the parable of the weeds to more fully understand God's actions on earth.

Matthew chapter 14 begins with Jesus leaving a home where he was staying and taking a seat near the shore of the Sea of Galilee. In ancient times teachers sat when they taught, rather than standing behind a podium. Tradition locates Jesus' discourse at the "Cove of the Parables," a horseshoe-shaped amphitheater where Jesus' voice would be carried to a crowd of hundreds. Because Jesus was teaching in a rural area of Galilee, his audience knew about farming practices to include the damage that weeds do to wheat crops.

Likely, Jesus offered the parable of the weeds in his second year of public ministry. By this time scribes and Pharisees started to push back at him. They wanted Jesus to explain his "supposed" authority to teach and to give them a sign to verify his Messianic identity. Initially, Jesus spoke the parable of the wheat and weeds to a crowd. Later, he interpreted it for his disciples, but he never interpreted it for the crowds of people who came to listen to his teachings.

Parable of Weeds in a Wheat Field

Here is Jesus' parable of the weeds:

> The kingdom of heaven may be compared to a man who sowed good seed in his field, but while his men were sleeping, his enemy came and sowed weeds among the wheat and went away. So when the plants came up and bore grain, then the weeds appeared also. And the servants of the master of the house came and said to him, 'Master, did you not sow good seed in your field? How then does it have weeds?'
>
> He said to them, 'An enemy has done this.'
>
> So the servants said to him, 'Then do you want us to go and gather them?'
>
> But he said, No, lest in gathering the weeds you root up the wheat along with them. Let both grow together until the harvest, and at harvest time I will tell the reapers, Gather the weeds first and bind them in bundles to be burned, but gather the wheat into my barn (Matthew 13:24-30 ESV).

Characteristics of Darnel

In the parable of the weeds, the weed was most likely darnel, also known as tares and poison ryegrass. Darnel originated in the Mediterranean region including the Middle East. Darnel is distributed widely throughout Israel to include extreme deserts. Like wheat, darnel is an annual plant; it grows in two-to-three-foot spikes. Again, like wheat, darnel spikes terminate in a grain head. The darnel grain head is longer than that of the wheat head and darnel lacks the fuzzy appearance of a wheat head. Initially, the darnel plant—stems, leaves and flowers—is green and looks like wheat.

When wheat and darnel are less developed, most farmers can't tell the difference between the two; however, as summer progresses, darnel plants turn black. The black color is a way to differentiate darnel from wheat which is creamy brown. Further, when both wheat and darnel are fully mature wheat stalks are taller than darnel.

Even a few grains of darnel negatively impact wheat crop quality. Darnel seeds are poisonous to people and livestock. Because darnel seeds are similar in appearance, size, and weight to wheat, it is difficult to separate or remove darnel seeds from wheat seeds. Often darnel seeds are contaminated with a variety of crop pests and diseases such as fungus. The best form of invasive species management is prevention, i.e., a farmer plants clean wheat seeds. Recently developed herbicides kill darnel; however, ancient Israelite farmers had no herbicides. They relied on sowing clean seeds to get a clean wheat crop.

Meaning of Parable

The dialogue between the farmer and his servants showed the farmer's knowledge of both wheat and darnel seeds and growth patterns of both. Probably, the farmer didn't purchase wheat seed for planting. Bought wheat seeds could be contaminated with darnel. Instead, the farmer used last year's wheat seed from his own fields. He and his servants knew that the wheat seeds they planted didn't contain darnel seeds. The only way fields could be contaminated with darnel was if the darnel seeds were deliberately planted.

When I first read that an enemy contaminated a farmer's wheat fields with weeds, I had a hard time believing the farmer's assertion that an enemy sowed weed in his field. The farmer sounded like he was making an excuse for careless planting of his fields. After investigating agriculture practices in first century Israel, I learned that jealous farmers often contaminated rivals' fields to discredit a rival farmer and to increase the value of wheat from their own fields.

Darnel
Wheat

****

Plant fertilizer contributes to plant growth and optimal fruit production

1. Name some bad fruit (sins) that we produce in our lives. If you need help, read Galatians 5:19-21.

2. What does the Bible consider good fruit production in our lives? If you need help with this list, read Galatians 5:22-23.

3. For plants, i.e., wheat, grapes, even weeds, to grow optimally, the soil conditions have to be right. Routinely, gardeners fertilize soil where plants grow. In contrast, often we neglect to fertilize the soil of our lives. What could or should we do to fertilize our walk with Christ?

****

Several verses after Jesus told the parable of the weeds, he provided its interpretation to disciples.

> The one who sows the good seed is the Son of Man. The field is the world, and the good seed is the sons of the kingdom. The weeds are the sons of the evil one, and the enemy who sowed them is the devil. The harvest is the end of the age, and the reapers are angels. Just as the weeds are gathered and burned with fire, so will it be at the end of the age. The Son of Man will send his angels, and they will gather out of his kingdom all causes of sin and all law-breakers, and throw them into the fiery furnace. In that place there will be weeping and gnashing of teeth. Then the righteous will shine like the sun in the kingdom of their Father (Matthew 13:35-43).

Jesus' interpretation affirms that at the final judgement wheat and weeds would be separated, i.e., the righteous placed in the barn or storehouse of God and the unrighteous thrown into a fiery furnace. Repeatedly, scripture writers affirmed that the course of human history is set toward judgement. No human will escape judgment.

Jesus described how weeds came into the world and why they were allowed to grow. The origin of weeds on Earth was Satan who sowed weed seeds in the pristine fields that God created. Remember, Satan sowed disobedience to God back in the Garden of Eden. Because God doesn't want anyone—any man or woman—to end up in the fiery furnace, he restrains his angels. They aren't allowed to pull out unrighteous, evil doers, until they mature. When it is abundantly clear that weeds are indeed weeds and not good wheat, angels reap or remove them, not to be placed in the storehouse of the Lord, but to be destroyed.

****

You have forsaken your first love (Revelation 2:3-4).

1. What should be a Christian's first love?

2. How should we make amendment for forgetting our first love? If you aren't sure, read Revelation 2:5.

3. If you repent, amend your life, and return to your first love, what will Christ do for you (Revelation 2:7)?

****

Outcome

In Revelation, John described a vision of the earth being harvested (Revelation 14:14-20). Grapes rather than weeds were harvested; however, the vision mirrors Jesus's parable of the weeds. Angels harvest grapes and throw them into the great winepress of God's wrath. The grapes were trampled and blood flowed outside the press. In this vision, John noted that the winepress was located outside the city. At the final harvest, evildoers aren't in the city wall or in God's barn (storehouse). In contrast, individuals who die in the Lord, i.e., they have a personal relationship with Christ, are blessed because their righteous deeds follow them into eternity (Revelation 13:13).

Takeaway

In my home garden, weeds keep growing back even though I pull them out. My husband tells me to spray them with weed killer. I don't because I am afraid the will killer will kill my flowers.

Chapter 13
LILY OF THE FIELD

Getting Started: Read Matthew 6:28-30

Jesus's parable of lilies adorning field grass is part of the Sermon on the Mount. In it, Jesus explained what he expected of members of his kingdom. This address is the standard for Christian life. The parable of the lilies of the field is about anxiety and worry.

****

Anxiety is uneasy
about something that may not come to pass.

1. Have you ever been anxious? What does that feel like to you? According to Proverbs 12:25, how does anxiety affect a person?

2. What we should do when we see others being anxious (Isaiah 35:4)?

3. How can you claim Isaiah's advice in your own life? Consider memorizing Isaiah 35:4 and repeating it out loud when you feel anxious).

****

Backstory

Before the Sermon on the Mount, Jesus was baptized by John. Following his baptism, Jesus fasted for 40 days alone in the wilderness.. After he heard that King Herod imprisoned John the Baptist, Jesus went to live in Capernaum on the northwest side of the Sea of Galilee. There, Jesus called several disciples and began his teaching and healing ministry.

On the northwestern corner of the Sea of Galilee near Capernaum is a gently sloping hillside, often considered the Sermon location. The hillside would have been covered with grass interspersed with wildflowers. Possibly, Jesus sat at the bottom of the hillside while listeners sat in the grass at higher elevations similar to an amphitheater. Matthew wrote that Jesus' disciples came to him, and he began to teach them. Most likely the "them" that Matthew identified referred to both Jesus' disciples and the crowd present.

In the Sermon on the Mount, Jesus included what we are blessed by experiencing. He spoke about the salt of the earth and warned that if the salt losses its flavor it will be good for nothing. Jesus gave God's standards for murder, adultery, divorce, and love of enemies.

Parable of Lilies in a Field

In the parable of the lilies, Jesus seems to chide listeners. His exact words were:

Jesus began this parable by asking listeners why they worried about clothes. In early Galilee, many individuals had no more than one-or- two sets of clothes. One set was for everyday wear and one for festive occasions. Jesus wasn't asking why individuals worried about *which* outfit they wore on a particular day. Instead he asked why they worried *if* they had clothes to wear.

Then, Jesus said that King Solomon in his splendor—in his rich, colorful garb—wasn't dressed as gloriously as a lily of the field, a simple field flower. In contrast to peasant's clothes made primarily from wool, goat hair, or camel chair, Solomon's clothes were made of linen, cotton, or silk. Often Solomon's outer mantle or robe was decorated with the fur of exotic animals. Solomon's clothes were usually purple, the symbol of royalty; however, on special occasions, the king wore white garments to denote purity, cleanliness, and joy. Peasant clothes were undyed or dyed brown.

****

How many outfits do you have in your closet?

1. If you stop buying clothes for yourself for a year, would you be more or less anxious?

2. Estimate the amount of extra money you would save.

3. Imagine giving that money to a fund that provided food and clothing to orphans or widows, i.e. , Orphan Medical Network International or Samaritan's Purse. Seriously, how do you think you would feel about donating this amount to a Christian charity?

****

Characteristics of Anemone (lily)

When Jesus used the lilies example, most likely he wasn't referring to the true lily described by the prophet Hosea. Instead, Jesus meant the anemone, a flower that grew wild and abundant in fields, was colorful, and known to most people. Often, anemones are called windflowers because even a gentle breeze will cause them to sway on their thin stalks. Anemones are perennials; they regrow year after year.

In Israel, the anemone blooms December through April. A single flower blooms on a leafless stalk. Normally, flowers have six petals. Most of these field flowers were red; however, anemone blossoms could be white, pink, purple, or blue. Flowers with red petals are most resistant to harsh growing conditions; they abound in Israel's steppes and deserts. Anemone flowers open in the morning and close in the evening. Although anemones have no nectar, flowers attract beetles and bees that feed on pollen.

Meaning of the Lily Parable

In the Sermon on the Mountain, the parable was about worry or anxiety. Three times in the Matthew 6:25-34 passage, Jesus told listeners not to worry. Specifically, Jesus identified four things they shouldn't worry about—their life, what they would eat, what they would drink, and what they would wear. Jesus promised that God knows his children's needs and God will meet those needs. Pragmatically, Jesus asked, "Who by worrying can add a single hour to his life" (Matthew 6:27)?

Stop reading right now! Go back and reread Jesus' question: can you add a single hour to your life, can you change a single outcome, can you increase or decrease what you have or don't have by worrying?

My aunt was a worrier, but she didn't just worry in her own mind, she worried out loud. We all heard her anxieties, her concerns, and even her dour predictions which worry generated. She lived a life of "what if." What if this happens, what if that happens. As a young girl, I didn't like to be around her. I didn't want to hear about all of the negative things that could happen. No, I'm not recommending that we keep our heads in the sand; however, remember the old saying that we don't need to borrow trouble by speculating about what could happen?

Anemone
Maria Lin

****

Who by worrying can add a single hour to his life (Matthew 6:27)?

1. By asking this question, what conclusion was Jesus trying to lead his listeners to?

2. How are you, more importantly, how could you apply the parable of the anemone to your life?

****

Outcome

Have you ever seen wildflowers growing in a meadow as you drove along a little-used road? If so, you may have wondered why all the beauty was there? Why did God put beautiful flowers in a field that so few people see? Perhaps the answer is that God made the beauty for just one human—you. On the other hand, perhaps God put the vibrant wildflowers in grassy meadows to remind himself that he knows every blade of grass, every flower, and every human

Living on Earth, Jesus' life was transient. As a human he touched few individuals in Palestine. Jesus gave beauty in that time which continued for 21 centuries. His beauty and lesson will continue and touch lives in the future however long that time will be.

Takeaway

From my perspective wildflowers are transient but they are anxious about their short life span. In God's timeframe, each person is transient. We have a choice in our transient life span; we can worry or have faith in God.

Chapter 14

GROWTH OF MUSTARD SEED

Getting Started: Read Mark 4:30-32

Jesus seemed to like the mustard seed. He used it several times to illustrate parables and teachings. The current parable focuses on the kingdom of God. This parable appeared in Matthew, Mark, and Luke. In Matthew it was placed between the parable of the weeds and Jesus' explanation of that parable.

Backstory

In African and Middle Eastern countries, the mustard tree (*Salvadora persica*) is called the toothbrush tree. Indigenous to Persia (Iran), the mustard tree could have entered Palestine along trade routes. Alternatively, returning Jewish exiles may have brought tree seeds back from Persia and planted them in gardens and fields.

Jesus was in Galilee when he told the parable of the mustard seed. Because Jesus grew up in a small village, he was aware of plants, their growth patterns, and their uses. He knew that mustard seeds are one of the smallest of all seeds; yet this tiny seed grows into a tree large enough for birds to perch on its branches.

Parable of a Mustard Seed

Mark record of Jesus' parable:

> What shall we say the kingdom of God is like, or what parable shall we use to describe it? It is like a mustard seed, which is the smallest seed you plant in the ground. Yet when planted, it grows and becomes the largest of all garden plants, with such big branches that the birds of the air can perch in its shade (Mark 4:30-32).

Characteristics of Mustard Seeds and Trees

In Jesus' time in Galilee, the mustard seed was *Salvadora persica.* We living in the west shouldn't confuse the seed and tree which Jesus referred to with the *Brassica nigra*, or mustard plant which grows readily in Bible gardens in the United States. The *Brassica nigra* grows to a maximum of 5-6 feet and is an annual.

The mustard tree is an evergreen shrub or tree that grows to 20 feet. Often mustard tree's many branches start low to the ground so that mustard trees grow as wide as tall. Mustard trees grow quickly and reach full size in a few years. The Middle Eastern mustard tree grows and thrives even in challenging environments where we would never plant a tree or crops.

Mustard trees were used for shade because of their low-growing branches. Unlike Jotham's thorn bush, the mustard tree had no thorns to deter people from resting beneath it. Although wild animals sometimes fed on tree shoots, many branches grew tall enough that predators couldn't reach birds that sat on them.

Mustard tree pods
Mustard tree
Maria Lin

Meaning of Mustard Seed Parable

The mustard tree's growth pattern illustrated how the Christian church would grow. Like the branches of a mustard tree, Christianity spread wide, throughout the Roman Empire, and into India and Ethiopia in the first several centuries after Christ's life and death. Similar to the fast growth of a mustard tree, Christianity flooded the ancient world. In a few centuries Christianity went from unheard of, to being outlawed, to the official religion of an empire.

Although the church began in a small province of the Roman Empire, it grew rapidly to become larger than the mightiest empire on earth. The parable was prophetic. In it, Jesus described what was to come.

****

Was growth of the Christian church inevitable?

1. What direction did Jesus give to his disciples to grow the church? If you aren't sure, read the following verses, then, answer the question.

    a. Mark 16:15

    b. Matthew 28:19

    c. Acts 1:8

2. What do you think Heaven will be like?

3. Is the Kingdom of God different from Heaven?

****

In Jesus' parable of the mustard tree, birds settled in its branches. Some Bible commentators took this clause as a warning to keep the early Christian church pure. In the ancient Near East, the phrase "birds of the air" was used to symbolize demonic forces in Scripture. When Jesus noted that birds settled in the mustard tree branches, he may have been warning disciples that Satan would attempt to encroach into the Kingdom of God.

Certainly, Jesus' warning became reality. Judaizers advocated that newly converted Christian could be justified only by observing Jewish laws, i.e., circumcision and adherence to dietary laws. Further, Gnostics claimed to possess an elevated knowledge, a "higher truth." The higher knowledge was acquired not from the Bible or the apostles' teachings, but from a higher mystical plain of existence. Individuals with this special knowledge, for example, Jezebel in the Thyatiran church, believed they were elevated above other Christians because of their deeper knowledge of God.

Many powerful men attempted to stop the initial spread of "The Way," the early name for the Christian church. The Jerusalem Jewish leadership tried to stop it when they arrested Jesus and turned him over to Pilate for crucifixion. Herod attempted to stifle the teaching of the apostles when he killed the apostle James and had Peter arrested. Paul was beaten, confined to house arrest, jailed, and finally murdered; however, his letters, many written when he was confined or jailed, formed the backbone of the spread of the good news of Jesus to the Gentile world. Jesus' parable of the mustard seed came true in the early centuries of the first millennia after Jesus's death.

****

A witnesses testimony for Jesus

Use Mark 16:15, Matthew 28:19, and Acts 1:8 and answer the next three questions:

1. It is the 21st century. Do Christians today need to spread the good news of Christ?

2. What is the difference between being a witness of Christ with your words and with your behavior?

3. There was an obituary about a woman that said she belong to x Christian church and was active in y and z activities in church. A colleague told me that she worked with the woman every day for 23 years and didn't know she was a Christian. How would you feel if someone said that about you?

****

Importantly, the parable of mustard seed remains true in the 21st century. Even with anti-Christian rhetoric and many churches turning from God's commands, Christianity continues to spread. The church Jesus founded is inevitable.

At this time the Christian faith seems to be spreading more in developing countries (Africa, South America, Asia), than in westernized countries. Important for Christians is the answer to the question, "Where will Christianity inevitably spread and grow even deeper roots?" Will it be in Greece, Rome, or Spain, where Christianity was first embraced. Perhaps, Germany and Great Britain, homes to great reformation thought? What about the United States founded on principles of religious liberty? We need to pray that all countries experience revival in Christian thought so more individuals can rest in the shade of God's love.

Outcome

Think back on the Egyptian, Greek, and Roman empires. Each culture had so many gods and goddesses that it was almost impossible to remember who was whom and who did what. Yet,

I don't know any stories where a god died for the sins of a believer, or individuals who weren't yet believers.

In contrast, the Christian religion at its inception and currently has one God. God's son came to live on Earth and then died for the sins of each man, woman, and child on Earth at that time and for those who would be born in the future. Like a mustard tree providing shade to people and livestock, Christianity provides shade and protection for every individual on earth.

Throughout the Roman Empire, men, women, and children were persecuted and murdered because they acknowledge Christ as their Lord. Have you ever wondered why first century Christians, who suffered so severely for their faith, still remained believers and refused to deny they were Christians?

I asked myself that question. Then, I recalled a recent television news clip in which 20 Christian men were beheaded with the beauty of the Mediterranean Sea as the backdrop. Their crime was belief in Jesus; they refuse to deny Jesus and convert to another religion. These men were no different than individual martyrs in first and second century Roman Empire.

The second-century Church father Tertullian wrote that "the blood of martyrs is the seed of the Church." Seeing a person stand firm in their faith despite slander, persecution, and even death often causes on-lookers to investigate that faith. Some come to a believing knowledge of Jesus. Thereby, more individuals rest under the canopy of the Christian church.

****

Will 21st century Christians be persecuted for their righteous actions?

1. Read Matthew 5:10. What did Jesus say about being persecuted for righteous behavior? Was Jesus ever persecuted for his righteous behavior?

2. Often, we don't recognize religious persecution. We believe that people are unkind because of some quirk in our personality. Don't be so quick to blame yourself when you are persecuted. Consider that you are being persecuted for your righteous behavior/standing for God.

3. Identify 2-3 situations where you were persecuted for your personal righteous behavior.

****

Takeaway

A former pastor believes that the 21st century is more like the first century church than any other century in time. The Roman Empire and western civilizations are similar. Like the first century Roman Empire, westernized society embraces varied religious beliefs along with situational ethics and values.

In the Roman Empire, Christian believers were under attack from non-believers and civil authorities alike. Today, claiming to be a Christian in western society causes ridicule, subtle, and, not so subtle, persecution. Often, twenty first century Christians are required to participate in actions that conflict with their faith, or just as egregious, keep silent.

Chapter 15

MOVING A MULBERRY TREE

Getting Started: Read Luke 17:6

In a parable about a mulberry tree, Jesus alluded to the power of faith. Although Matthew, Mark, and Luke, recorded several of Jesus' teachings on faith, this one-verse parable was recorded only by Luke.

Backstory

When Jesus spoke the parable of the mulberry tree, he was talking privately to his disciples. Have you ever wondered why Jesus used a mulberry tree to illustrate the outcome of a small amount of consistent faith? If I wanted to illustrate the role of faith to move a tree, I would have used the largest tree I knew of, i.e., a cedar or the Tabor oak. Most likely, Jesus chose a mulberry tree because one was nearby. Jesus said, "this mulberry tree," from which we can surmise that Jesus and disciples were sitting under a mulberry tree or could see one from where they were seated.

Jesus outlined responsibilities for Christian discipleship:

1. Never lead a person into sin. Jesus said it would be better for a disciple to be thrown into the sea with a millstone tied around his neck than to cause another to sin.

2. Rebuke a brother if he sins.

3. Forgive a brother if he repents of sin and asks for forgiveness even as often as seven times a day. Seven times a day meant that the repentance-forgiveness cycle was limitless.

After hearing these requirements, disciples were overwhelmed with the responsibilities of followership. How could they live so blameless a life that they never caused another individual to sin? Did they have the courage to rebuke a fellow Christian when he or she sinned? How could they, simple men that they were, forgive and forgive and forgive?

A Parable on Faith

In desperation, the disciples cried out to Jesus, "Increase our faith!" (Luke17:5). The disciples wanted greater faith to live up to the standards Jesus set.

Jesus gave them a one sentence response that was a parable:

> If you have faith as small as a mustard seed, you can say to this mulberry tree, 'Be uprooted and planted in the sea,' and it will obey you (Luke 17:6).

The Greek word for faith used in this parable is *pistis*, which means moral conviction and reliance on Christ for salvation. Also, p*istis* has a more abstract definition which includes "consistency." By using *pistis,* Jesus told disciples they could meet the responsibilities that he outlined through consistent reliance on him.

****

Now faith is being sure of what we hope for and certain of what we do not see (Hebrew 11:10).

1. Write down your definition of faith— that you have, not an idealized version.

2. Look up the following verses about faith:

   a. Romans 1:17

   b. 2 Corinthians 5:7

   c. Hebrews 11:6

   d. 1Peter 1:7-9

3. Rewrite your definition of faith. How did it change from question #1 to question #3?

****

Characteristics of Mulberry Trees

The mulberry tree is also called the black mulberry and sycamine tree. In Israel, mulberry tree remains were found that dated from the late Iron Age. Mulberry trees are popular in areas with long, hot summers to include areas of drought. The mulberry tree can reach a height of 33 feet, but most times it is shorter. It has a short, stout trunk, and a broad canopy.

Mulberry trees grew throughout Judea and Galilee. Finding Jesus and disciples sitting under a mulberry tree is consistent with the tree's characteristics and growth pattern. During our stay at Kibbutz Lotan in the southern Negev Desert area, we saw a 25-foot-tall mulberry tree which was planted for shade.

Unlike many plant flowers, mulberry tree flowers are one sex or bisexual; sometimes flowers change from one sex to another. Mulberry trees were domesticated for their fruit; yet trees don't bear much fruit before they are 15 years-of-age. Once fruiting

Black Mulberry
Mulberry tree
Maria Lin

starts, trees produces fruit for hundreds of years. Frequently, older mulberry trees produce better tasting fruit than younger trees. In the same way that a mulberry tree produces better tasting fruit with age, so Christians are often more spiritually productive after several (or many) years of studying God's word. The mulberry tree being the last tree to bud in the spring. Because the mulberry waits until all possibility of frost is past, it is called the "wisest of all trees."

Meaning of the Mulberry Tree Parable

When his disciples asked Jesus to increase their faith, his response was that they could and would accomplish great tasks with a consistent small amount of faith in him. At that point in their walk with Jesus, probably the disciples didn't have unswerving faith in him. They were still getting to know Jesus and coming to realize that he was the promised Messiah. Jesus was wise enough to meet the disciples where they were in their awareness of him.

****

Most of us have faith in something.

1. Name some things or beings that you have faith in.

2. How do you exhibit this faith? In other words, how do you act on your faith in these things or people?

3. How is having faith in the person of Jesus different from having faith in your church or your church doctrine?

****

Because mulberry tree flowers bloom last in the spring doesn't mean that trees bloom late. Mulberry flowers bloom at just the right time to be fertile, pollinated, and produce mulberries each year. Have you ever noticed how God seems to time everything, i.e., events, relationships, etc. perfectly?

Back when Jesus told his disciples if they have as much faith as a mustard seed, they could move a mulberry tree, in essence Jesus was telling them: "You only need a little faith in me now." What he didn't tell them was that later, when he was arrested, crucified, and murdered, then was the time that they needed a lot of faith in him. A minister once told his congregation that "God is rarely early, but he is never late."

****

Does size matter? Is amount important?

My husband and I have a silly joke. When I say to him that I am sorry for something, he asks me how much? Just sorry; very sorry; or very, very sorry?

1. If you have to differentiate between a little faith and a lot of faith, how would you go about making that measurement?

2. Should you wait until your faith in Jesus is large, or fully mature, until you attempt to do something for Jesus, that is, you step out in faith?

3. How is it possible to have a lot of faith on one topic i.e., believe that Christ is the Savior of the world, while having only a little faith on another topic, i.e., if I tithe on my income before taxes, I will have enough money to pay my bills.

****

Outcome

Jesus was captured, placed on trial, and crucified until he died. Bible description suggested that not one apostle had sufficient faith in Jesus as Messiah to stand with him during his trial. Fear overcame faith. At that time, not one apostle had sufficient faith to move a mulberry tree.

Thankfully, apostles' behavior during Jesus' trial, death and burial wasn't the end of their faith/faithless story. After his resurrection, Jesus showed that he forgave each apostle their faithlessness and restored them to full fellowship with him.

Each became on fire for Jesus and the new religion based on his being and teachings. By the fourth century, Christianity was the official religion of the Roman Empire. At the present time there are over one billion Christians. Christianity is the largest religion on Earth.

Takeaway

Reliance on Jesus means to be dependent because you have confidence (faith) based on experience. Reliance on Jesus doesn't come easy in today's society. Like the apostles, our reliance in Jesus occurs over time. Reading the Bible and implementing its guidelines for living over years is required.

Chapter 16

UNSHAKABLE REEDS

Getting Started: Read Luke 7:18-28

John the Baptist's public ministry lasted only one to two years. Then, Herod Antipas arrested him. John was imprisoned at Machaerus, a fortress for political prisoners. While there, John sent several disciples to Jesus. They asked Jesus if he was the promised Messiah. After they left, Jesus contrasted John's firm convictions with reeds blown by the wind.

****

Think about your ideas about convictions.

1. What is a conviction? Look up the word in a dictionary. List three of your convictions.

a.

b.

c.

2. Is it possible to have a conviction that you don't hold to firmly? In the above list, highlight or check the ones that you don't firmly hold to.

3. Look at the remaining convictions. If you believe them and hold firmly to them, how do they impact your life?

****

Backstory

Christian tradition teaches that John the Baptist was the son of John and Elizabeth; therefore, John the Baptist was from a priestly family. Jesus and John were cousins although probably not first cousins.

Jesus went to John to be baptized. Although this baptism was the start of Jesus public ministry, it was the high point of John's. Soon afterward, King Herod arrested John. The reason for the arrest was John's criticism of Herod for divorcing his powerful Nabatean queen to marry his sister-in-law, Herodias. Herod was indignant that John had the audacity to hold him—a king—to a moral standard.

The Bible recorded that King Herod used this personal reason for imprisoning John, however, getting John out of circulation made political sense. When John preached that individuals who came to him for baptism must change their behavior. John's demanded changes impacted King Herod's income. For example, John told tax collectors to collect only the amount of money required by Rome. They should stop lining their pockets and those of King Herod by overtaxing citizens. Likewise, John taught that soldiers must be content with their pay and stop extorting money from people. John had tremendous influence with people in Herod's kingdom. According to Josephus, Herod feared that John, with his widespread support from the common people, would instigate rebellion against him.

****

Compare and contrast John the Baptist and Jesus.

1. Consider how the two of them dressed, where they lived, who their friends were, the manner in which they spoke, etc.

2. Compare the behaviors of the civil and religious authorities toward each man.

3. Today, in America how do you think that we would treat John the Baptist? How would we treat Christ if he lived today?

****

After John was in prison perhaps 15-18 months, he sent two disciples to Jesus. They asked Jesus if he was the expected Messiah or if they should wait for someone else (Luke 7:19). Jesus didn't give the disciples a direct "Yes" or "No" answer. Instead, Jesus told the disciples to go back to John and report what they saw and heard, i.e., the blind received their sight, the lame walked, and lepers were cured.

The Parable of Unshakable Reeds

After John's messengers left, Jesus asked the crowd what they expected when they went to see John in the desert. In his questioning Jesus told a parable in which he contrasted John the Baptist's behavior with a reed that blew in the wind, swaying first one way then another. Here is how Luke recorded Jesus parable:

> What did you go out into the desert to see? A reed swayed by the wind? If not, what did you go out to see? A man dressed in fine clothes? No, those who wear expensive clothes and indulge in luxury are in palaces. But what did you go out to see? A prophet? Yes, I tell you, and more than a prophet. This is the one about whom it is written:" 'I will send my messenger ahead of you, who will prepare your way before you.' I tell you,

among those born of women there is no one greater than John; yet the one who is least in the kingdom of God is greater than he (Luke 7:24-28).

Characteristics of Israeli Reeds

The reed that Jesus referred to when talking about John was the Israeli reed, sometimes called the giant reed and Cypress cane. Reed colonies were located on the banks of natural water courses, in floodplains of medium or large sized streams, and in dry riverbanks far from permanent water sources. Reeds grew throughout Israel from Mount Hermon to the Negev Desert. They lined the Jordan River

The giant reed is a perennial; it comes up year after year. Reeds can grow to 20 feet and may grow 10-12 feet in a single growing season. The central reed stalk is called a culm; culms are about one and a half inches in diameter and hollow. Each culm has leaves that resemble those on corn stalks; however, leaves have sharp edges that can cut fingers. As drier weather prevailed, foliage turned light brown and rattled in the wind. Giant reeds bend with the wind, even when they grow in large colonies. In ancient times, and even today, reeds were used as wind breaks and to stop soil erosion.

Individuals who came to John for baptism knew about reeds; they saw reeds growing along the Jordan River. To them reeds elicited mainly positive thoughts. Probably, they remembered how Isaiah associated reeds with humility. Jesus denied that John was similar to a swaying reed; instead, John was firm and upright. Jesus averred that John's beliefs were certain, and John lived by them. John stayed on message (repentance) and on task, i.e., baptism. John didn't have a politically correct bone in his body. He didn't pander to public opinion, giving one message to common people and a second one to the rich

Israeli Reed
Maria Lin

and powerful. John called the Jerusalem elite “a brood of vipers.

Meaning of Unshakable Reed Parable

Although Jesus denied that John was a swaying reed, John’s purpose reflected how reeds were used in ancient Judea and Galilee. By his words and life, John stood against the erosion of godly behavior. He called ordinary citizens, tax collectors, and civil and religious leaders to change their lives to reflect God’s standards. Also, like reeds used as windbreaks, John stood as a buffer between people who were righteous and the secular society of the Roman Empire. The best windbreaks lower wind chill in man, animals, and plants. Everything we know about John the Baptist showed a priest and prophet who lived close to God. As a windbreak, John lowered the chilling effects of a secular Roman society on inhabitants of Galilee and Judea.

Outcome

Less than a year after John the Baptist disciples visited Jesus and this parable King Herod murdered John by cutting off his head. Jesus parable included a eulogy for John. Jesus eulogized John by saying of all the men (humankind) born of woman, there was none greater than John the Baptist. John was no weak reed, or as we would say in modern times, “no shrinking violet.”

Takeaway

What do you want in your eulogy?

Think about several characteristics you want to be remembered for, i.e., you were a woman of character, you were a man who loved his family, you were loyal, etc. Now that you have identified what you want to be remembered for, what do you think will really be said in your eulogy? Starting today, what can you change about yourself to make your eulogy be what you would like it to be?

Chapter 17

OLIVE OIL FOR LAMPS

Getting Started: Read Matthew 25:1-13

The parable of ten virgins emphasizes the importance of being prepared for Jesus' return to Earth. Preparation encompasses planning! Last spring, I attended a "Prepper's Convention." Vendors were selling everything imaginable to prepare buyers for natural or manmade disasters, i.e., dried food, batteries, small stoves, guns, and ammunition. Seeing the hundreds of participants caused me to think about preparing for Jesus' second coming. If all convention attendees could prepare for disasters, surely I could prepare for the greatest future event ever to come, the second coming of Jesus.

****

Christians must prepare their minds for action!

When Peter exhorted individuals in the scattered Christian church to prepare their minds for action, he wasn't recommending a contemplative lifestyle, attending multiple Bible study, or only listening to Christian radio.

1. Read 1 Peter 1:13-16, what did Peter expect Christians to do?

a.
b.

c.
d.

2. Are you doing what Peter exhorted you to do?

3. Are there areas in your life that you want to change, i.e., need correction?

****

Backstory

Olive oil had a central place in Israelite history. In the Tabernacle, olive oil was the fuel for the seven Lamps on the Golden Candlestick. Olive oil was used to anoint and set apart kings, priest, and prophets. Olive oil was an ingredient in the Presence Bread.

The parable of virgins with their lamps used imagery from wedding customs of first century Palestine. The wedding feast took place in the evening; the day was filled with dancing and entertainment. At day's end, the bridegroom along with his good friends went to the home of the bride's parents. The bride left her parent's home and joined her bridegroom. They headed to the groom's home to a banquet held in their honor. Friends joined the groom and bride along the way to their home. Often, the procession was halted along the way for the couple to receive congratulations or share a toast. Eventually, the bride and groom reached the groom's home and enter it. The door was closed; the wedding feast began. No one else was permitted entry.

Some Bibles translate the 10 virgins as bridesmaids who were typically girls younger than the bride. Perhaps, they were 10-11 years old. These bridesmaids for some reason weren't a part of the groom and bride's procession. Rather, they waited in the streets or outside the groom's home. The bridal party was delayed; it didn't arrive until around midnight.

Lamps such as the bridesmaids used required large amounts of oil to keep them burning. Depending on the size of the oil lamp, olive oil had to be replenished as often as every 15 minutes.

Parable of Olive Oil Lamps

Jesus told the parable of the 10 virgins to his disciples. At the time Jesus was seated on the Mount of Olives. It was Passover week. In a few days, Jerusalem's Jewish leaders would capture Jesus, turn him over to Pilate, and demand that Romans crucify him. Here is the parable:

> The kingdom of heaven will be like ten virgins who took their lamps and went out to meet the bridegroom. Five of them were foolish and five were wise. The foolish ones took their lamps but did not take any oil with them. The wise, however, took oil in jars along with their lamps.
>
> The bridegroom was a long time in coming, and they all became drowsy and fell asleep. At midnight the cry rang out: "Here's the bridegroom! Come out to meet him!" Then all the virgins woke up and trimmed their lamps. The foolish ones said to the wise, "Give us some of your oil; our lamps are going out."
>
> "No," they replied, "there may not be enough for both us and you. Instead, go to those who sell oil and buy some for yourselves."
>
> But while they were on their way to buy the oil, the bridegroom arrived. The virgins who were ready went in with him to the wedding banquet. And the door was shut. Later the others also came. "Sir! Sir!" they said. "Open the door for us!"

> But he replied, "I tell you the truth, I don't know you." Therefore keep watch, because you do not know the day or the hour (Matthew 25:1-13).

Characteristics of Olive Trees and Oil

The source of the oil identified in the parable was the olive tree. Olive oil fueled the lamps of the waiting virgins. The olive tree is one of the world's oldest cultivated trees. The olive tree is indigenous to present day Israel and Syria. Its dry composition makes the olive tree drought tolerant. Olive trees live up to 1000 years. Very old trees can be hollow, secondary to fungus having attack the wood. Leaves appear gray-green on top, while the bottom has a silvery appearance.

The olive is oval shaped and contains a seed. Usually, unripe olives are green and ripe olives are purple-black. Between October and January olives are harvested in Israel. The most common time in Jerusalem is October to November. Ancient peoples used long sticks to shake olive tree branches with the result that ripe olives fell to the ground. Ripe olives were gathered. Olives allocated for oil were put into an olive press.

Ancient olive presses were heavy hewed stones with a vertical stone wheel. The wheel was rotated by an animal (donkey or ox) or by persons. The oil and watery liquid squeezed from the pulp was separated in settling vats. Water sank to the bottom while oil floated on the top. Oil was drained through a linen cloth to remove impurities.

In first century Palestine, olive oil was used for healing, i.e., the Good Samaritan used olive oil to heal the wounds of the man who was robbed and beaten on the way to Jericho.

oil Lamp
Olive tree
Maria Lin

****

The Holy Spirit illuminates a Christian's life.

1. The Hebrew word for olive tree, olive, and olive oil was *zayith* which means yielding illuminating oil. Read, ponder, and discuss how the Holy Spirit illuminates or should illuminate your life. Some Bible verses that will aid you are:

   a. John 14:16-18
   b. John 16:8-13
   c. Hebrews 10:16

2. Identify several ways you allow or will allow the Holy Spirit to illuminate your life.

3. What parts of your life are you hiding from the Holy Spirit and not allowing him to touch?

****

Meaning

In this parable the bridesmaids were central figures; however, they were not central to the meaning of the parable. Nor was their number (10) important. Because the parable identified an equal number of wise and foolish virgins doesn't mean that there will be an equal number of righteous and unrighteous individuals on Earth at Jesus' second coming. Likely, Jesus gave the figure to round out the story.

In the Bible, the bridegroom is synonymous with Jesus, the Christ. Often the bride is the Christian church, composed of all who believe Jesus is the son of God and acknowledge him as Lord of their life. In the early Christian church, olive oil symbolized the illuminating presence of the Holy Spirit.

Through the Holy Spirit's presence, Christian actions shines in a dark world, healing the world.

The interpretation of the parable of the oil lamps is that when Jesus comes the second time, it will be too late to rethink our lives or initiate a "do over." At this time, Jesus' followers will enter into the wedding banquet with the bridegroom. They will have eternal life with God. The unprepared and foolish, who rejected or ignored Jesus, will be shut out; they will endure eternal punishment.

Importantly, the parable wasn't about Jesus' first coming to Earth. It's about his return to Earth the second time. In the parable Jesus continued his assertion from Matthew chapter 24 that only God knows when Jesus' second coming will occur. The parable is one of three, i.e., parable of the ten virgins, talents, and the sheep and goats, with the same message. All emphasized the truth that Jesus will come a second time and the importance of being spiritually prepared.

****

Scout motto is "Be prepared"

Prepared means to plan in advance, make ready beforehand, and to work out the details.

1. That Scout motto means that Scouts must be ready, willing, and able to do what is necessary in any situation that comes along, even in an emergency. Prepared means to plan in advance, make ready beforehand, and to work out the details. How can you use the Scout motto in your life as you wait for Jesus' second coming?

2. What do you have to plan so you will be ready for Jesus' second arrival? Alternatively, are you right where you should be and need no different preparation?

3. Probably, Scouts prepare for an emergency by taking a cardiopulmonary resuscitation (CPR) class. List changes or actions you need to take to be optimally prepared for Jesus' second coming.

a.

b.

c.

****

Outcome

In first century only a minority of Jews looked for the promised Messiah. Even fewer understood that Jesus, the Messiah, would come a second time. Generally, Jesus' apostles believed that his second coming would be soon after his ascension, even within their lifetimes. But, decades passed, decades turned into centuries, and centuries into millennia.

For us living in the 21st century, it is easy to doubt that Jesus will come again. It's easy to fall asleep waiting for Jesus' second coming. It is easy to be unprepared, to stop planning ahead, so that if Jesus' return is delayed (by our standards of promptness), we are caught unaware and unprepared. When Jesus returns, individual preparedness will count. Preparedness can't be shared or transferred just as the wise virgins couldn't share their oil with the foolish ones.

Takeaway

The prophet Malachi's point was that fearing and serving God and looking for the Messiah wasn't futile. These activities had long-term rewards. Similarly, for Christians living a life consistent with God's laws and statutes now will have long-term rewards. Typically, we name this parable the Parable of the Wise and Foolish Virgins. That name doesn't reflect the interpretation of the parable.

Chapter 18

WILD OLIVE BRANCH

Getting Started: Read Romans 11:16-24

Despite Jewish Christians starting the Christian Church in Rome, Gentile Christians resisted accepting Jewish Christians into their fellowship. Paul's letter to the Romans (about 71 AD) included a parable using cultivated olive and wild olive trees to illustrate Gentile's proper response to their Jewish Christian brethren.

Backstory

Paul's purpose in writing the parable of the in-grafted wild olive branch was to remind Gentile believers that the root of the Christian faith was in God's covenants with the Jews. God promised to bless all nations through Abraham's seed. When he reminded the Rome church about God's promises, Paul's wanted to encourage a fully integrated church. Paul wasn't attempting to make the Christian church in Rome a sect of Judaism, nor was he advocating that Gentile Christians replace the Jews in God's favor.

Initially, the church in Rome was composed of Jews who believed that Jesus was the Messiah. The Church in Rome wasn't started by an apostle, but by Jews who returned from Pentecost in Jerusalem. Almost immediately, Jewish believers evangelized Gentiles. Then, Emperor Claudius banished all Jews from Rome. For 12 years the Christian church in Rome consisted of only Gentiles.

When Nero became Emperor, he invited Jews back to Rome, noting that they were good for business and trade. The problem was that Gentiles refused to allow Jewish Christians back into the Christian church in Rome. Gentile Christians may have concluded that Emperor Claudius' rejection of the Jewish Christians meant that God rejected them. Because Rome was the capital city of the Roman Empire, this discriminatory attitude had the potential to spread beyond Rome.

****

Imagine yourself as Jewish Christian who started the Christian church in Rome.

1. Ponder your feelings when you return to your home church in Rome after 12 years of forced exile only to find that you weren't welcome. Identify how you would have felt.

2. In this scenario, how would you feel about the present Christian church in Rome? Would you be emotionally tied to them or feel you have a vested interest in the church you started?

3. How could Paul's parable in Romans 11 make you feel more secure?

****

Paul focused his evangelistic efforts on the Gentiles. In Paul's letter to the church in Rome, Paul made it clear that the current parable was designed for Gentile believers. He wrote it to counter Christian Gentiles' arrogant belief that they were better than Jewish Christian believers. Bases for Gentile arrogance was that unlike Jews, Gentiles didn't reject Jesus and lobby for his crucifixion. Further, Gentiles never denied that Jesus rose from the dead as many Jerusalem Jewish leaders denied the resurrection.

Parable of Wild Olive Branches

Here is the parable of grafting wild olive branches onto a cultivated olive tree that Paul wrote to the Romans:

> If the root is holy, so are the branches. If some of the branches have been broken off, and you, though a wild olive shoot, have been grafted in among the others and now share in the nourishing sap from the olive root, do not boast over those branches. If you do, consider this: You do not support the root, but the root supports you. You will say then, "Branches were broken off so that I could be grafted in." Granted. But they were broken off because of unbelief, and you stand by faith. Do not be arrogant, but be afraid. For if God did not spare the natural branches, he will not spare you either.
>
> Consider therefore the kindness and sternness of God: sternness to those who fell, but kindness to you, provided that you continue in his kindness. Otherwise, you also will be cut off. And if they do not persist in unbelief, they will be grafted in, for God is able to graft them in again. After all, if you were cut out of an olive tree that is wild by nature, and contrary to nature were grafted into a cultivated olive tree, how much more readily will these, the natural branches, be grafted into their own olive tree! (Romans 11:16-24).

Characteristics of Wild Olive Branches

Common names for the wild olive tree are oleaster and Russian olive. Wild olive trees are often 12-15 feet tall, so they are more often a shrub than tree. Wild olive trees are multi-trunked and often grow as wide as tall. Wild olive trees grow almost everywhere, i.e., in all types of soil and water. The oleaster is drought tolerant and indifferent to wind and heat.

The wild olive tree has deep taproot (central root) and well-developed lateral roots. The wild olive spread roots horizontally in search of nutrients. This root system adds to the stability of the wild olive tree. Like the cultivated olive tree, the wild olive produces a drupe-like fruit; however, fruit is smaller than and not as tasty as olives from cultivated olive trees. In most countries, wild olives aren't eaten. They may be used to make olive oil.

Meaning of Wild Olive Branch

In the parable of the in-grafted wild olive branch, Paul identified

1. a root and branches (boughs) of a cultivated olive tree
2. a branch (bough) of a wild olive tree
3. grafting a wild olive branch onto a cultivated olive tree.

The original cultivated olive tree with its root and branches is the promises God made to Abraham, Isaac, Jacob, and their offspring, the Jews. This root was solid and sure. Jews who believed in Jesus as Messiah were the original root and branches of the early Christian church. Branches broken off from the cultivated olive tree were Jews who refused to believe that Jesus was the long looked for Messiah. This was the majority of Jews who lived in Palestine and throughout the Roman Empire at the time.

The wild olive branch equaled Gentile Christians who believed that Jesus was the son of God and followed his teachings. The interpretation of Paul's parable in Romans chapter 11 is that the Gentile believers were grafted into—became an integral, productive off shoot—of the Jewish faith.

Although Paul identified that a wild olive branch was grafted onto the root of the cultivated olive tree, olive growers rarely graft wild olive branches onto cultivated olive trees. In nature, just the opposite occurs: growers graft cultivated olive tree branches onto wild olive tree roots. Their desire was to increase the cultivated olive crop. Paul was aware of this normal grafting procedure. Paul wrote that his parable was contrary to nature. Perhaps, Paul believed that making his point was more important than technical accuracy about olive tree grafting.

Horticulturists identify three reasons for tree grafting:

1. To propagate trees that don't root well by cutting a shoot (small branch) from the poorly growing tree and grafting it onto a healthy tree

2. To obtain a stronger root system

3. To grow plants faster

Importantly, the root (Judaism) sustains the newly grafted branch (Christianity). The newly grafted branch (Christianity) doesn't sustain the root (Judaism).

In Paul's parable, all three reasons for grafting support grafting the newly converted Gentile believers into roots of Judaism. Gentiles used the structures and traditions of the established Jewish faith as roots for their worship of Jesus. An example is the Jewish tradition of meeting weekly to hear and study God's word. Using this Jewish tradition, new Christian church members fellowshipped regularly and became more knowledgeable about their faith. Further, Jews had sacred God-inspired writings. Knowing about and hearing Jewish scriptures.

Wild olive tree branch
Maria Lin

(Old Testament) facilitated more ready acceptance of the New Testament gospel and letters. In these ways, the Christian faith grew stronger and faster and became a deeper part of Gentile-convert's lives. The new Gentile believers and churches acquired spiritual richness and fertility by being grafted into the deeply rooted, cultivated olive tree (Jewish faith).

****

In your church are some individuals more important than others?

Using Paul's carefully written words in the entire Romans 11 chapter and particularly the parable of the in-grafted wild olive branch, answer the following questions.

1. What groups are more important than other groups in your church?

2. What groups are you in? Describe, discuss, and explain your conclusions. As you write or verbalize your answer, forget about a "socially acceptable" response, or even a "Christian" response. Write/verbalize your true feelings.

3. What do you think Paul would conclude about your church from your answers?

****

Despite the seeming lesser value of wild olive tree products than cultivated olive tree products, Paul's parable didn't mean that Jewish Christians were more valuable than Gentile Christians.

Outcome

The growth of the new Christian church occurred almost everywhere in the Roman Empire. The Christian church appeared and thrived even in the most unhospitable environments. Growth patterns mirrored the growth of wild olive trees which grow many places.

Approximately two-thirds of the U. S. population self-identifies as Christian. Like the wild olive tree, many individuals have a root deep in their Christian faith. At the same time individuals aren't necessarily tied to one church denomination. They spread horizontal roots in search of an optimal church family.

Takeaway

The wild olive branch will never produce the same type of olive that grows on a cultivated olive tree. Similarly, Christianity is a unique religion and doesn't produce the same fruit as Judaism. In this parable, the in-grafted wild olive tree branch resonates with most of us. We are the wild olive branch grafted into the cultivated olive tree. Gentile believers form most of the body of believers in churches in westernized countries.

Chapter 19

THORN-INFESTED LAND

Getting Started: Read Hebrews 6:7-8

The Hebrew's author inserted a two-verse parable. He believed that an agricultural parable about land producing crops versus thorns and thistles was an illustration that readers in the first century could understand. He emphasized that thorn-infested (unproductive)land had no worth.

Backstory

Hebrews was addressed to Jewish Christians, but, was relevant to Gentile Christians as well. For centuries biblical scholars believed that Paul wrote the letter to the Hebrews. Within the past 500 years other writers, i.e., Barnabas and Apollos, were proposed. Who wrote Hebrews isn't as important as its message.

Immediately, before the parable of productive versus thorn-infested land, the Hebrew's author wrote some hotly contested statements. The writer asked: can an individual who has rejected Christ, after he has been enlightened and shared in the blessings of the Holy Spirit, be brought back to repentance? The writer averred that by their action individuals who have fallen away crucify Jesus over again and subject Jesus to public disgrace. In the Bible, Annias and Sapphira were a couple who accepted all the benefits of Christianity but later denied Christ by their actions.

****

Can a born-again Christian become thorn-infested?

1. How can you differentiate between good land and thorn-infested land? What about a good and bad person?

2. Can thorn-infested land produce good crops? How do bad people produce good fruit?

3. What will God do to land that doesn't produce good crops, i.e., that is thorn-infested? What should a Christian do to a purported Christian that doesn't produce good crops?

****

Parable of Thorn-Infested Land.

The Hebrews writer give the parable of productive versus thorn-infested land.

> Land that drinks in the rain often falling on it and that produces a crop useful to those for whom it is farmed receives the blessing of God. But land that produces thorns and thistles is worthless and is in danger of being cursed. In the end it will be burned. (Hebrews 6:7-8).

Characteristics of Thorn-Infested Land

The Mediterranean buckthorn is a slow growing thorn common in the Mediterranean Basin where Jewish converts lived in the first century Christian church. The buckthorn is unattractive and doesn't normally grow in cultivated gardens or fields. It prefers to grow in poor soil that is gritty and highly eroded. Along with thistles, the buckthorn is the last species to disappear when livestock over-graze an area.

Buckthorn is tangled with many branches. Grayish stems are topped with thorny spikes. Small flowers are yellowish and inconspicuous. Initially green, fruit turns black when mature. Although birds eat the fruit, humans find it bitter. To humans fruit acts as an emetic and laxative. In large quantities, fruits are toxic to humans

Aphids are attracted to the Mediterranean buckthorn. If the buckthorn grows in a damp climate, it tends to develop a fungal disease. Once aphids and fungus appear on plants, they often spread to more valuable plants in the area.

The parable writer suggestion that land with buckthorn should be burned is an excellent one. One Hebrew word for thorn is *shayith* which translates as scrub, thorn, or trash. Trash is worth little or nothing. Buckthorn should be removed and destroyed (burned) so it doesn't regrow somewhere else.

****

The Hebrew word for thorn means trash!

1. What would it take for a fallen individual to re-establish a relationship with Christ?

2. What is the likelihood that a person will re-establish a relationship with Christ after falling away? Can it be done?

3. If you have a loved one who falls away from belief in Christ, what can, or should, you do to try to restore them to Christ?

Buckthorn branches
Buckthorn shrub
Maria Lin

Meaning of Thorn-Infested Land Parable

In the parable, the writer may be answering his own question of whether any individual can receive the benefits of being a Christian then by his/her actions deny Christ.

Notice that the subject of the parable is land—not rain, not a farming process, not a crop, not a farmer—but land. The land receives rain and does something with the rain. Notice, in this parable rain comes on both good land and on thorn infested land.

In both cases, the land produces something. In one instance the land responds by producing a useful crop. In other words, the land produced grains, trees, herbs, etc. that gave a farmer food for his family. Possibly, the crop was abundant enough so that some food could be sold, thereby enhancing food security for an entire community. In contrast, other land responds to the rain by producing thorns and thistles. This land is worthless to the farmer. He curses this land. In the end it will be burned.

****

What is the effect of rain on a thorn-infested field?

1. Will increasing the amount of rain that falls on the thorn-infested land cause the land to produce good crops? What happens when thorn-infested land receives more rain?

2. What happens to thorn-infested land if a farmer adds fertilizer to the soil? Will fertilizer on the thorn-infected land cause good crops to grow there?

3. Think of yourself. As a Christian you are productive land; you are no longer thorn-infested land. How do you demonstrate that you are productive land?

Outcome

Trash is an excellent symbol for men and women who learned what Christ did for them, tasted the heavenly gift and goodness of the Word of God, shared in the Holy Spirit, and then turned away from the goodness of God. The outcome for these individuals isn't the storehouse of God, but a burning trash heap.

We aren't in Ananias and Sapphira's minds; consequently, we don't know if they were truly converted to Christianity. Acts gives the impression that they completed the Christian "To Do" list, i.e., profession of faith in Christ as Savior, public baptism, indwelling of the Holy Spirit, taking communion (body and blood of Christ), church attendance; but perhaps they didn't have a personal relationship with Christ. Alternatively, perhaps this early church couple moved deliberately away from Christ. Worldly events became more attractive than Christ. Whatever happened in this couple's lives, we don't want the same thing to happen to us. We don't want to crucify Christ anew or subject him to public disgrace. We don't want to die unrepentant.

Similarly, church-attending, seeming Christian individuals with negative thoughts or wrong doctrine are thorn-infested. They can spread their beliefs to the community of believers. hey are thorn-infested land.

Takeaway

Church leaders need to identify an optimal way to remove, or negate the effects of buckthorn, from their congregations. They can't allow false doctrine to continue in their congregation.

Chapter 20

WORMWOOD, ENDTIMES

Getting Started: Read Revelation 8:10-11

The parable and event described is part of the endtime Bible literature. When an angel sounded a third trumpet, a star, blazing like a torch, fell to Earth. The blazing star was named Wormwood after a bitter plant by the same name (Wormwood) that grew on Earth. The parable is about bitterness and death.

The Hebrew word wormwood (*laanah*) is used about eight times in the Old Testament. In the Old Testament, "wormwood" is synonymous with a bitter taste or poison. It was used as a metaphor for Israelites' idolatry, calamity and sorrow, and false judgments.

Backstory

When the first trumpet sounded, a mixture of hail, fire, and blood was hurled down onto Earth. The result was a major ecological catastrophe—one third of trees and one third of green grasses were scorched. When an angel sounded the second trumpet, something like a huge blazing mountain was thrown into the sea. The result was that one third of the sea creatures died and one third of ships were destroyed.

****

Bitterness is harmful or destructive to our mind, emotions, and will.

1. Your will encompasses your disposition, inclinations, passions, appetite, even your desires. It's how you customarily act. How can your will be influenced and poisoned by bitterness?

2. If you don't like or want bitterness in your life, what strategies can you implement to remove it?

3. Identify at least one strategy that you will start implementing today. At the end of the day, check how you did, using this strategy to impact your bitterness. Be honest.

****

Readers of Hebrews were familiar with another story of bitter water. Most knew every step of the Israelites exodus from Egypt. They remembered that three days after the Israelites left Egypt they arrived at the spring at Marah in the Desert of Shur. At Marah, the Israelites expected to find fresh water; instead, the water was bitter and undrinkable.

The Bible didn't identify what made the Marah spring water undrinkable; however, Moses action provided a clue. God told Moses to throw a piece of wood into the spring. Jewish rabbis believe the wood was from a mangrove tree, one of the few trees that leaches saline from water. The result was that the Marah spring water became drinkable. The reverse of the miracle at Marah happens in John's parable in Revelation.

Wormwood Parable

The sounding of the third trumpet was significant in that for the first time John recorded that people died from the effect of the trumpet sounding. Here is the parable:

> The third angel sounded his trumpet, and a great star, blazing like a torch, fell from the sky on a third of the rivers and on the springs of water—the name of the star is Wormwood. A third of the waters turned bitter, and many people died from the waters that had become bitter (Revelation 8:10-11).

Characteristics of Wormwood Plant

Wormwood is an herb, native to the eastern Mediterranean area to include Canaan/Palestine. With the exception of rue, wormwood was the bitterest herb known in ancient times.

The plant wormwood regrows year after year; it is a perennial. Ancient people believed that if wormwood was hung to dry, it couldn't be placed in full sun. or its aromatic properties would be lost. In contrast to many plants that thrive in sunny locations, wormwood prefers to grow in the shade. The shade-loving properties of common wormwood echo the pall and moral decay endemic in ancient Israel and Judah before they were conquered.

Meaning of Wormwood Catastrophe

21st century readers can relate to John's vision of a star falling to Earth and contaminating one third of the Earth's fresh water. The result will be death and subsequent bitterness in the hearts of individuals who live.

Imagine the effects of the first three judgements on the Earth: one third of trees and plants destroyed, scarce food, and maritime trade curtailed with destruction of one-third of the

Wormwood
Maria Lin

Earth's ships. Many items from foreign countries will be unavailable i.e., spices, technological items, fruits, and vegetables. Oil imported to heat homes, fuel vehicles, and support manufacturing will be reduced.. Finally, water from faucets may be undrinkable. Family and neighbors will die. People will be sorrowful and bitter over loss of life and loss of their livelihood. Our spirits will be overwhelmed.

Recipients of John's letter understood the significance of one third of the fresh water turning bitter. Mankind needs fresh water to live. Although a person can live 21 days without food, few can live more than 3-5 days without water when the temperature is about 70-degree Fahrenheit. In tropical areas where temperatures are substantially higher or in extremely dry climates like in the Middle East, individuals survive less time without fresh water. Thirsty people will do one of two things: die from dehydration or drink the contaminated water and die from its poison.

Individuals appreciate how this falling star can be named after a bitter, poisonous plant. The meaning is in the effect of eating the plant and consuming the outcome of the shooting star, Woodworm. Both result in bitterness, through tasting the plant and through experiencing death of loved ones.

Earth, isn't alone in the cosmos. Celestial events have, do, and will impact our planet and lives. When we read about a star falling to Earth, we think of a meteor (shooting stars) or comets that enter Earth's atmosphere and sometimes landing on Earth. If the celestial traveler doesn't burn up as it passes through Earth's atmosphere, the falling star is large. Wormwood will be huge.

The question is how Wormwood that falls in a specific location will cause widespread death and bitterness, likely around the globe. Scientist have offered at least two explanations:

1. The cosmic traveler contains contaminated particles released as it moves through the Earth's atmosphere. When contaminated particles fall to Earth, they poison fresh water.

2. When the meteor lands on the Earth's surface, it sends dust and high levels of pollutants, they rise into the atmosphere. With the rotation of the Earth and wind patterns, polluted particles spread around the globe. When contaminants return to the Earth's surface, they contaminate rivers, lakes, streams, and other fresh water used for drinking.

****

What do you think about the Wormwood parable?

1. What are the chances that a meteor will fall to Earth? Could the Bible description of a meteor be some type of bomb?

2. Will there be outcomes such as spread of contaminants around the Earth?

3. If the parable occurs, do you want to be alive after the fall of Wormwood? Do you think you will be alive.

****

Outcomes

Wormwood parable is futuristic. God's foremost reason for causing Wormwood to fall on Earth, consistent with all his actions, is to get individuals attention and cause them to repent. The book of Revelation provides little information about the number and characteristics of individuals who repent as a result of this futuristic action.

Another perspective is that God is using the outcomes of blowing the seven trumpets to destroy the present Earth to prepare it for the new Heaven on Earth. This parallel is similar to razing a house so a new and better structure can be built on the spot.

Takeaway

When the seven seals are opened and the trumpets sound millions of Earth's people will die. Despite the severity of God's judgements, most of humankind refuses to repent of their sins and acknowledge God as sovereign. The Bible says that people will continue to worship demons and idols.

REFERENCES

1. Strong, James. *The New Strong's exhaustive concordance of the Bible.* Nashville, TN: Thomas Nelson Publishers, 2010.

2. HaReubeni, Nogah and Helen Frenkley. *Tree and shrub in our biblical heritage*. Kiryat Ono, Israel: Neot Kedumim Limited, 1989.

3. Modzelevich, Martha. *Flowers in Israel.* http://www.flowersinisrael.com, 2016.

4. Adeyemo, Tokunbo, ed. *Africa Bible commentary*. Grand Rapids, MI: Zondervan, 2006.

5. MacDonald, William. *Believer's Bible commentary*. Nashville, TN: Thomas Nelson Publishers, 1995.

6. Lockyer, Herbert. *All the parables in the Bible*. Grand Rapids, MI: Zondervan, 1963.

7. Borowski, Oded. *Agriculture in iron age Israel*. Winona Lake, IN: Eisenbrauns, 2009.

8. Whiston, William. *The Works of Josephus, Complete and Unabridged*. Peabody, MA: Hendrickson Publishers Inc., 1987.

9. Millay. Edna St. Vincent. *First fig. A few figs from thistles*. New York, NY: Harper & Brothers, 1922.

10. Rabinowitz, Louis I. *Torah and flora*. New York, NY: Sanhedrin Press, 1977.

11. Faith and Health Connection, *Teaching biblical truths for health and wellness*. http://www.faithandhealthconnection.org, 2016.

ABOUT THE AUTHOR

Dr. Carolyn A. Roth is a spiritual woman who believes in God the Father, God the Son, and God the Holy Spirit. She believes that words in the Bible are God-inspired. She is an ordained minister.

Carolyn has advanced degrees in psychology and a doctorate from the University of San Francisco. She spent 25 years teaching in universities. Her last job was as a distinguished professor.

She initiated and designed the award-winning Bible Garden in Virginia. Carolyn is a Master Gardener. Four of her other books are about plants in the Bible.

She writes for laity rather than clergy or theologians. Some of her books, like this one, are designed as Bible studies.

Carolyn is eager to hear your questions and comments. Contact her at carolyn.roth@ymail.com. This book is available on Amazon in paperback, electronic and audio formats.

ADDITIONAL BOOKS BY AUTHOR

Bible Plants Series

Gardening Genesis, Meditations to live by.

Good News in Nature, Examining plants in the Gospels.

Rooted in God, Interpreting plants in Bible lore.

Living it up Series

Level Up, Time to mature your faith.

Power Up, Strategies for women to make a difference in their churches.

Spiritual Living Series

Connecting the Church Calendar, 101 meditations on church seasons.

Lesser-Known Bible Characters, Using their relationships to restore ours.

Lore, Legends, Lies, Separating Bible truth from deceit.

Out of the Shadows, Exploring the power of Bible women.

Made in the USA
Monee, IL
03 November 2026

67384784R00098